国际五校研究生联合设计课程
INTERNATIONAL JOINT STUDIO FOR FIVE UNIVERSITIES

融合·共生与介入·整合
——华侨大学厦门校区与村落的城市设计研究

MERGING AND SYMBIOSIS & INTERVENTION AND INTEGRATION
STUDY ON URBAN DESIGN OF HUAQIAO UNIVERSITY XIAMEN CAMPUS AND VILLAGES

刘塨 许懋彦 张路峰 邓蜀阳 古谷诚章 主编

中国建筑工业出版社

图书在版编目（CIP）数据

融合·共生与介入·整合——华侨大学厦门校区与村落的城市设计研究 / 刘塨等主编． — 北京：中国建筑工业出版社，2018.9
ISBN 978-7-112-22641-2

Ⅰ．①融… Ⅱ．①刘… Ⅲ．①高等学校－校园－建筑 设计－作品集－中国－现代 Ⅳ．① TU244.3

中国版本图书馆CIP数据核字（2018）第 203794 号

本书是以清华大学、华侨大学为主等五所国际型大学的研究生联合发起的设计课程项目为主题，国际五校联合的课程使得不同地域、不同文化的学生聚集一起，针对同一设计命题提出自己的设计理念，通过文化的碰撞交融促成新颖的设计理念产生，进而产生出优秀的设计作品。书中收录的即是以此联合设计的优秀作品，给建筑学、城市设计等相关专业学生以借鉴。

责任编辑：杨　晓　唐　旭
责任校对：王　瑞

国际五校研究生联合设计课程
融合·共生与介入·整合——华侨大学厦门校区与村落的城市设计研究
刘塨　许懋彦　张路峰　邓蜀阳　古谷诚章　主编
*
中国建筑工业出版社出版、发行（北京海淀三里河路9号）
各地新华书店、建筑书店经销
北京富诚彩色印刷有限公司印刷
*
开本：880×1230 毫米　1/16　印张：17 $\frac{1}{2}$　字数：828 千字
2019 年 1 月第一版　　2019 年 1 月第一次印刷
定价：228.00 元
ISBN 978-7-112-22641-2
　　（32773）
版权所有　翻印必究
如有印装质量问题，可寄本社退换
（邮政编码　100037）

序 Preface

多校联合设计已经成为国内外建筑教育的热点，其同题共进及工作坊互动等教学模式，对于共享优质教学资源、博采各教学体系之长以及对建筑设计教学机制的深度挖掘与响应、建筑教育的国际化等方面，有极其重要的意义。

2016～2017年，由清华大学、早稻田大学、重庆大学、中国科学院大学、华侨大学五所高校建筑学专业研究生，针对华侨大学厦门校区与周边城中村的现状及发展前景，分别以"融合与共生——华侨大学厦门校区校园空间可持续发展概念设计"、"整合与介入——基于校村共生的华侨文化保护与侨村环境整治方案设计"为题，连续进行两次多校联合设计教学，取得丰硕成果。在此将学生作业成果及配套教学资料整理出版，以求交流分享并获得各位同仁的批评指正。

我们的联合设计教学有三个特点：

一是对多校联合教学模式的完善。

通过"Fast-Meeting"系统，搭建了远程双向互动的视频教学平台，突破多校联合教学的时空瓶颈，使其互动环节从短暂的现场工作坊模式扩展到一个大作业的全过程。选题的内容时间和要求都有了充分发挥的自由，充分体现了"多校联合"的特征。

二是对"研究性设计"模式的完善。

国内建筑学专业教育，本科多以设计毕业，研究生多以研究论文毕业，"研究性设计"就成了研究生培养中的关键性转折环节，我们针对性地安排了四个阶段：

1. 预研阶段——开题及相关理论文献资料学习；
2. 现场阶段——调研发现并提出问题；
3. 方案阶段——根据问题提出解决策略与方法；
4. 深化阶段——整合策略与方法形成设计成果。

整个过程，始终强调由发现问题到解决问题的逻辑过程的连续性。总体过程体现清晰的研究逻辑与方法，再把成果表现为图式化的思维与综合性的设计。

三是在人居环境热点领域进行学术性选题。

当今快速城市化取得了巨大成就，但也给我们的人居环境带来巨大的挑战。城市空间扩张速度远远超出社会发展节奏，政府主导的基础设施网络急剧扩张，切割原有村落肌体，土地征收与房地产开发使得大量传统村落迅速消失、衰落。土地投机的随意性与拆迁补偿的复杂性，造成"城中村"等城乡交错的普遍现象。华侨大学厦门校区现状及其过程就是这种现象的一个典型案例。

厦门校区的空间网络是十年左右形成的，周边村落的空间网络则是百年或更长时间形成的，两者在时间上存在巨大冲突；当下的城市化，超出城市自我有机更新的节奏——把自身的增长融合到自己的有机脉络之中的那种固有的节奏，这正是传统社区的特点。

厦门校区是现代大学以公共属性为基础的大尺度空间网络，周边村落的空间网络是传统的以私密属性为基础的小尺度空间网络，两者在空间上也存在巨大冲突。

这种时空的截然对立，使两者成为彼此的异质斑块而不相容，成为城市人居环境脉络的巨大创伤。在"新常态"背景下，在新型城镇化前瞻视野中，如何化解这种对立和矛盾，如何通过"融合与共生"的策略实现人居环境生态回归和谐，实现由量的扩张到内涵提升的转变，这是时代对我们提出的挑战，也是本课题的核心内涵。

这是一个非常有特色的实际个案，但却体现当下具有极大普遍性的学术问题。这样我们就有了一个坚实的研究起点。同时，作为个案又有鲜明的个性，现有的研究成果无法直接套用，必须结合实际创造性应用，这又会催生另外一个层面的学术问题，这样我们就获得研究的动力。

我们把课题定位于城市设计，侧重概念设计，要求在整体空间格局、分区组团结构及典型建筑单体三个层面做出连续性逻辑概念设计表达。

两次课题，我们安排了两个不同的逻辑进路。2016年我们的题目是"融合与共生——华侨大学厦门校区校园空间可持续发展概念设计"。强调自上而下的思考，具有形态学特征。2017年我们的题目是"整合与介入——基于校村共生的华侨文化保护与侨村环境整治方案设计"，强调自下而上的思考，具有发生学的特征。两者结合成为完整的思路。

同学们很好地完成了我们的教学要求。来自不同学校的同学，甚至同一学校的不同课题小组都体现出鲜明的特色。统一的题目、统一的教学要求，并没有限制他们的思路，反而让他们思考得更深入、更广阔，产生很多令人惊喜的创新之举。

这些充满活力的作业成果表明，同学们深刻体会到城市问题不是单纯的建筑学问题，而首先是一个社会学问题，由此他们获得了开阔的视野，在社会学的大背景下展开建筑学的思考。有的从自然生态与社会生态的统一去寻求融合与共生；有的把"华侨文化"作为学校与村落的社会公约数来寻求两者的融合；有的从"校园空间社会化"与"乡土空间现代化"去实现双向对接等，产生许多很有启发、很有学术价值的成果。有些直接转化为他们毕业论文的选题。

还有一个令人称道的成果，那就是这些成果的总结，在学术上解决了华侨大学厦门校区在十年规划建设中始终没有解决的规划理念体系的统一与完备，对于承办方这是一个偌大的收获。希望这些成果对大家也有启发作用。

五校联合设计教学团队
2018年5月

五校联合设计"全家福"
Five Universities Design Hotchpotch

2016年国际五校研究生联合设计全体成员合影

2017 年国际五校研究生联合设计全体成员合影

教师简介
Teacher Introduction

清华大学

许懋彦 课程主持教授

清华大学 建筑学院 教授、博士生导师

弋念祖 课程助教

清华大学 建筑学院 博士生

早稻田大学

古谷诚章 课程主持教授

早稻田大学 创造理工学部建筑学科 教授

山田浩史 课程助教

早稻田大学 创造理工学部建筑学科 讲师

王薪鹏 课程助教

早稻田大学 创造理工学部建筑学科 博士生

张卓群 课程助教

早稻田大学 创造理工学部建筑学科 博士生

中国科学院大学

张路峰 课程主持教授

中国科学院大学 建筑中心 教授、博士生导师

彭相国 课程助教

中国科学院大学建筑中心 讲师

王大伟 课程助教

中国科学院大学建筑中心 讲师

程力真 课程助教

中国科学院大学建筑中心 博士生

重庆大学

邓蜀阳 课程主持教授

重庆大学 建筑城规学院 教授、博士生导师

阎波 课程主持教授

重庆大学 建筑城规学院 教授、博士生导师

翁季 课程主持教授

重庆大学 建筑城规学院 教授、博士生导师

华侨大学

刘塨 课程主持教授

华侨大学 副校长 建筑学院 教授

郑志 课程主持教授

华侨大学 建筑学院 教授

姚敏峰 课程助教

华侨大学 建筑学院 副教授

赖世贤 课程助教

华侨大学 建筑学院 讲师

目录 Contents

序　　　　　　　　　　Preface

教师简介　　　　　　　Teacher Introduction

上篇：融合与共生——华侨大学厦门校区校园空间可持续发展概念设计
Merging and Symbiosis: Conceptual Design for Spatial Sustainable Development of Huaqiao University Xiamen Campus

2016 国际五校研究生联合设计教学计划
2016 International Joint Studio for Five Universities Teaching Plan 2

清华大学	Tsinghua University	5
早稻田大学	Waseda University	29
重庆大学	Chongqing University	35
中国科学院大学	University of Chinese Academy of Sciences	68
华侨大学	Huaqiao University	98

2016 国际五校研究生联合设计过程回顾
2016 International Joint Studio for Five Universities Review 122

2016 国际五校研究生联合设计教学总结会会议记录
2016 International Joint Studio for Five Universities Teaching Plan 126

下篇：介入与整合——基于校村共生的华侨文化保护与侨村环境整治方案设计
Intervention and Integration: Preservation and Revitalization Design for the Traditional Village-Within-The Campus in Huaqiao University

2017 国际五校研究生联合设计教学计划
2017 International Joint Studio for Five Universities Teaching Plan 131

清华大学	Tsinghua University	134
早稻田大学	Waseda University	162
重庆大学	Chongqing University	168
中国科学院大学	University of Chinese Academy of Sciences	206
华侨大学	Huaqiao University	237

2017 国际五校研究生联合设计过程回顾
2017 International Joint Studio for Five Universities Review 267

2017 国际五校研究生联合设计教学总结与反思
2017 International Joint Studio for Five Universities Teaching 271

后记　　　　　　　　Postscript 272

2016 国际五校研究生联合设计课程
2016 INTERNATIONAL JOINT STUDIO FOR FIVE UNIVERSITIES

上篇
PART ONE

融合与共生

华侨大学厦门校区校园空间可持续发展概念设计
MERGING AND SYMBIOSIS
CONCEPTUAL DESIGN FOR SPATIAL SUSTAINABLE DEVELOPMENT OF HUAQIAO UNIVERSITY XIAMEN CAMPUS

2016 国际五校研究生联合设计教学计划
2016 International Joint Studio for Five Universities Teaching Plan

一、课题背景

在中国前阶段快速城市化的进程中，空间扩张速度远远超出社会发展节奏，政府主导的基础设施网络急剧扩张切割原有村落肌理，土地征收与房地产开发使得大量传统村落迅速消失、衰落。土地投机的随意性与拆迁补偿的复杂性，造成"城中村"等城乡交错的普遍现象。

与此同时，一线城市利用高校大扩招，纷纷建设大学城、文教区。这是我国高等教育由精英教育发展进入大众教育的重要阶段，给各高校带来巨大发展资源。厦门校区建设使学校发展空间翻倍，在学生规模扩大60%的前提下，彻底解决严重超标的拥挤不堪。同时师资引进、生源提高等方面获得巨大动力。大学城、文教区加速了城市化，但遗患匪浅。宏观上巨大的斑块尺度和相对单一的功能，造成严重生态失衡，将长时间游离于城市生态律动与整合。微观上又造成村校混杂交错的"村中校、校中村"的复杂状况。二者空间脉络上相互排斥，但经济上又相互依存：村中的"学生街"为学生提供日常生活服务，学校的物业、保安、保洁岗位为村民就业提供了机会，形成了村民与教师共跳广场舞的局面。如何将二者的关系从功能到形态进行整合，建立起和谐的融合与共生关系，是一个巨大的挑战。"新型城镇化"以及"新常态"思维的转变为下一轮城市化与城市有机更新提供了新的理论依据，为自然生态和社会生态的同步可持续发展、为医治城市化创伤、实现城市与社会的内涵发展提供了难得的机遇。

二、校园概况

华侨大学1960年创办，直属国侨办，是一所面向海外、面向港澳台，为侨服务，为经济建设与社会发展服务，传播中华文化的国家重点建设的综合性大学。现有全日制在校生28000余人，平均分布在泉州、厦门两个校区。

华侨大学厦门校区位于集美文教区，集美大道北侧，背靠天马山，面对杏林湖。规划总用地2000亩，总建筑面积63万平方米，"十一五"规划学生数15000人，远期20000人。2004年11月开工建设，2006年10月投入使用，现完成规划面积1300亩，建筑面积55万平方米，全日制在校生14000人。

校园规划突出生态理念，一是主体行为生态，把学生生活区布置在中间，到各点的距离在400米左右，故全校园以步行为主，骑自行车者不足3%。二是环境行为生态，突出体现在水环境，打通校园与周边的生态脉络，形成基于"功能湿地"的生态校园，不仅实现海绵城市功能，还实现污水零排放、节水1500吨/日、水禽保护地等众多生态效益。

建设基地原属兑山村，分为潘涂、下蔡、宅内、祖厝后、西头、西珩、南尾井七个自然村落。根据国侨办和厦门市政府的战略合作协议，拆迁由当地政府承担。拆迁现状：红线内剩下场地东南的潘涂、宅内、下蔡，约600农户，西北的南尾井、西珩，约100农户。校园中间还有一户"钉子户"，是典型的村校错落状态。

根据校园规划，华侨大学厦门校区分为教学区、生活区、文博区与运动区四个分区。教学区、生活区基本建设完成，运动区部分完成，"文博区"尚未开始建设。本课题所研究的范围即是校园规划中"文博区"的用地范围。根据原有校园规划方案，"文博区"拟有如下项目：

廖承志音乐厅，1500座；
大学生活动中心，15000平方米；
华侨博物馆，9000平方米；
校史馆，5000平方米；
档案馆，3000平方米；
其他配套内容，灵活考虑。

三、课题释义

本次课题的核心议题是探讨校园空间与社会空间的关系问题。校园和社会（村落）是两张不同的社区（Community）网络，二者分别具有不同的功能结构和物质形态，具有不同的构成规则和空间要素，以及不同的历史进程与演化规律。当二者"碰撞"在一起时，在不选择"大拆大建"即"一个吞噬另一个"的前提下，是否存在另一种可能性：二者通过融合得以共生？所谓"融合"是指空间形态上二者和谐共处（Spatial Integration），而"共生"是指在功能上互动互惠（Functional Inter-Activation）。如果要实现二者的融合与共生，二者的哪些空间要素和功能要素需要做出改变？哪些要素必须保持不变并且能够传承下去？

本课题拟定研究重点是：原规划中"文博区"的校园功能空间与现有自然村落如何以及在多大程度上能够实现融合与共生，即如何在总体和单体层面上，将原规划中"文博区"所设定的校园功能创造性地融入现有村落社会和空间结构之中，使村落生活与校园生活互利互动，有机融合。

四、教学目标

引导学生思考在新型城镇化的大背景下新的城乡关系以及城乡共赢的城镇化模式；

引导学生理解"生态校园"的可持续规划理念，理解"自然生态"与"社会生态"可持续发展的意义；

训练学生从城市尺度理解校园与村落空间，学习以微观的空间操作手法介入和干预一个复杂社会空间的方法和技能；

要求学生掌握文献调研与现场调研的方法。

五、教学阶段与任务描述

本学期教学分四阶段进行，各阶段任务与成果要求如下：

第一阶段 本阶段为"预研究"（Pre-Research）阶段，（《预研究专题举例》附后，供各校参考）要求各校内小组合作进行现场研究回顾与现状分析，文献研究与案例分析，理论阅读，提出总体设计策略与目标。

第二阶段 本阶段为"现场研究"（Site Survey）阶段。该阶段要求集中在现场进行实地调研和踏勘，五校师生混合编组，分不同专项进行调研工作（调研提纲小组自行拟定）。

第三阶段 本阶段为"概念设计"（Conceptual Design）阶段。各校内小组合作针对整个研究地块提出总体性、概念性规划设计方案。

第四阶段 本阶段为"深化设计"（Design Development）阶段。小组深化总体设计方案；个人选择与总体设计意图相关的单体、群体、空间节点或系统进行个体介入设计，完成设计成果并进行联合答辩。

第五阶段 各校分别整理完善成果，准备出版展览文件，校内答辩展示交流。

附：《预研究专题举例》(仅供各校参考，可根据各自理解添加或更改)
1. 城市化（城镇化）概念与中国的城镇体系
2. 中国的土地制度变迁：城市与乡村
3. 村庄产业结构、社会结构与空间形态的演变：传统与现代
4. "城中村"现象、成因与对策
5. 华侨文化与侨村的历史现状及未来
6. 闽南地区传统民居建筑类型
7. 地方政府土地流转与拆迁安置补偿相关政策
8. 中国高等教育发展与大学校园规划理念的演变
9. 中国的"大学城"：校园与社会
10. 生态社区与生态校园：观念与技术
11. 城市历史保护与活化再生策略与方法

1.Background

In the process of rapid urbanization in China, the space expansion speed is far beyond the pace of social development, and the original village body is cut by the sharp infrastructure network expansion led by the government, and land expropriation and real estate development has made a large number of traditional villages rapidly disappear and decline. In addition, randomness of land speculation and complexity of demolition compensation has caused the general urban-rural crisscross phenomenon, like village in city.

Meanwhile, the city constructs the university city as well as culture and education area using the large enrollment expansion of universities, which is an important stage of higher education from elite education to mass education, bringing huge development resources each university.

Construction of Xiamen campus of Huaqiao University doubles the school development space, with 60% scale of student expansion, solving serious crowding and providing great motivation of teacher introduction and enrollment enhancement. The university city as well as culture and education area have accelerated the urbanization, with deep problems left. Macroscopically, the huge plaque scale and single function cause serious ecological imbalance with ecological rhythm and integration for a long time. Microscopically, it causes the staggered village – within – campus and campus – within –village. There is space conflict and economy interdependence between them: the student street in the village can provide daily life services for students, and property, security and cleaning in the university can provide the employment opportunity for villagers, forming a situation of joint square dance of villagers and teachers. It is a huge challenge to integrate the relationship between them from function to form, to build a harmonious relationship of integration and symbiosis.

The change of "New-type Urbanization" and "New Normal State" thinking offers a new theoretical basis for the next round of urbanization and urban regeneration. Furthermore, it provides a rare opportunity for the sustainable development of natural ecology and social ecology, the healing of urbanization trauma and the realization of the connotation development of city and society.

2.Campus Overview

Huaqiao University, founded in 1960, directly under the Overseas Chinese Affairs Office of the State Council, facing overseas, Hong Kong, Macao and Taiwan, servicing for overseas Chinese and economic construction and social development, and spreading Chinese culture, is a key comprehensive university. There are 28000 full-time students in Quanzhou campus and Xiamen campus on average.

Xiamen campus, located in Jimei Culture and Education District and north of Jimei Avenue, backs against Tianma Mountain and faces Xinglin Lake. The total planned land is 2000 mu, total construction area is 63 square meters, number of students of the eleventh five-year plan is 15000 and 20000 at a specified future date. Construction began in November 2004, which was put into use in October 2006, and now the completed planning area is 1300 mu, construction area is 55 square meters, and the number of full-time students is 14000.

The campus planning can highlight the ecological idea: first, the main body behavior ecology: the student living area is set in the middle away from each point for 400 meters, so people give priority to walk, and cyclists are less than 3%; second, environment behavior ecology: in water environment, the campus and the surrounding have a connecting ecological context, so the ecological campus is formed based on the function wetland, not only realizing the sponge urban functions, but also realizing wastewater zero discharge, water saving of 1500 ton/ day, waterfowl protection area, and many other ecological benefits.

Construction base originally belongs to the Duishan Village, including Pantu, Xiacai, Zhainei, Zucuotou, Xitou, Xiyan and Nanweijing. According to the strategic cooperation agreement of the Overseas Chinese Affairs Office of the State Council and the Xiamen Municipal Government, the demolition shall be borne by the local government. For the status quo of demolition, there are southeast Pantu, Xiacai and Zhainei with about 600 farmers, northwestern Xiyan and Nanweijing with about 100 farmers, and middle Zucuotou within the red line, as a typical village – within – the campus.

According to the campus planning, Xiamen Campus is divided into 4 districts which are teaching district, living district, culture district and sports district. The construction of teaching district and living district are almost completed, while the sports district is partially completed but the construction of culture district has not yet started. The range of this research site is the same as the culture district. According to the original campus planning, the culture district will have the following items:

Liao Chengzhi Concert Hall, 1500 seats;
College Students' Activity Center, 15000 square meters;
Overseas Chinese Museum, 9000 square meters;
School History Museum, 5000 square meters;
Archives, 3000 square meters;
and supporting facilities required.

3. Topic Definition

The core issue of this subject is to discuss the relationship between campus space and social space. Campus and society (villages) are two different community networks, which have different functional structures and physical forms, different composition rules and spatial elements, and different historical processes and evolution rules. When the two "collide" together, is there another possibility that they can co-exist through fusion without choosing "mass-demolishing mass-construction", namely "one devouring the other"? The so-called "fusion" refers to the harmonious coexistence of the two in space and the "symbiosis" refers to the mutual interaction in function. If the fusion and symbiosis of the two are to be realized, which spatial and functional elements of them need to be changed? What elements must remain unchanged and be passed on?

This topic proposed study is focused on: the original planning "wenbo area" function of the campus space with the existing natural villages how and to what extent can realize the integration and symbiosis, i.e., how the overall and individual level, set the "wenbo area" of the original planning of campus function creatively integrate into existing village social and spatial structure, make the village life and campus life mutually beneficial interaction, the organic integration.

4. Instructional Objective

Guiding students to think about the win-win pattern of urbanization and the new relationship between urban and rural areas on the background of new urbanization.

Guiding students to understand the sustainable planning concept of "ecological campus" and the meaning of sustainable development on "natural ecology" and "social ecology".

Training students to understand the space of campus and village from the urban scale, and to learn the skill of intervening in a complex social space by using the micro spatial manipulation .

Requiring students to master the method of literature research and field research.

5. Teaching Stage and Task Description

Stage 1 This is "Pre-research" stage Teams in each school should cooperate in conducting site survey review and current status analysis, document research and case study, theory reading and putting forward the overall design strategy and objectives;

Stage 2 This is "Site Survey" stage Students should do field investigation and exploration on the site. Teams should be comprised by 5 students from different schools, and do research works on different monographic subjects;

Stage 3 This is "conceptual design" stage Teams in each school should propose their conceptual planning and design;

Stage 4 This is "design development" stage Teams in each school should deepen the overall design. Each team member should design individually, choosing one typical aspect including the monomer, group, space joint or planning system which is related to the overall design, and complete the final design production for joint reply;

Stage 5 Teams in each school should optimize the design results, prepare documents for publishing exhibitions, and then present the exchanges on campus.

清华大学
TSINGHUA UNIVERSITY

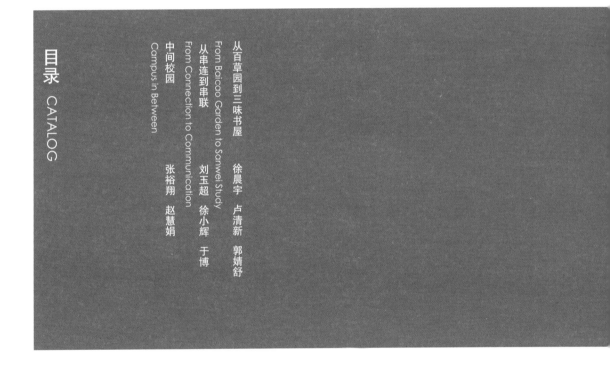

目录 CATALOG

从百草园到三味书屋　徐晨宇　卢清新　郭婧舒
From Baicao Garden to Sanwei Study

从串连到串联　刘玉超　徐小辉　于博
From Connection to Communication

中间校园　张裕翔　赵慧娟
Campus in Between

学生团队 Student Team

 赵慧娟

 张裕翔

 于博

 徐小辉

 徐晨宇

 卢清新

 刘玉超

 郭婧舒

上篇：融合与共生 —— 华侨大学厦门校区校园空间可持续发展概念设计

中间校园
CAMPUS IN BETWEEN

指导老师：许懋彦　课程助教：弋念祖
设计者：张裕翔　赵慧娟

场地问题 |

公共空间的缺失、建筑尺度的悬殊和道路阻断都导致了场地功能和活动的分离，进而造成了不同活动群体之间的隔离。

① 公共空间缺失

消极的景观处理　　停车挤占公共空间　　农田侵占公共空间

③ 尺度悬殊

校园尺度　　村落尺度

② 道路阻断

④ 功能分离

自发形成的非正式接口位置　　　　地段内主要功能分布现状

应对策略 |

针对地段中存在的主要问题，我们提出了相应的空间改造态度——"中间校园"。通过"中间式"选点形成空间的主干结构，并用"中间式"的内容策划方法，依据现有情况充实功能。最终使大学和城中村完成从建筑空间到现实活动的充分融合。

中间城市概念　　201台选址说明图

③ 尺度调和

建筑尺度过渡示意图
场地选定过程

空间策略

空间整合　　空间连接　　尺度过渡　　功能融合

①② 空间整合与连接

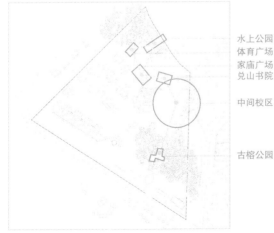

水上公园
体育广场
家庙广场
兑山书院
中间校区
古榕公园

"中间校园"空间结构图

④ 功能融合

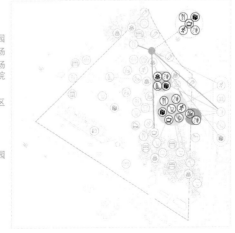

竞技运动
学术交流
商业消费
体育健身
文化探索

人群 – 活动分析图

在不同人群分享相同活动的前提下，通过为活动提供空间来促进人群之间的融合。

吸纳整合场地现有功能示意图

教师评语：

校与村的二元并置引发了对"边界"的再思考。边界可以是隔绝的屏障，也可以是沟通的桥梁。在围绕校园边界的观察中，作品将这一堵薄薄的围墙，扩大成校与村联系的"中间地带"。在现有界墙两侧已形成的功能互交通联系的基础上，作品通过对现有零散场地的再梳理、对过渡地带的再引导、对人群活动的再组织，力图结合建筑聚落的规划设计形成校村融合的新地带。

"中间校园"规划设计 I

"中间校园"总平面图

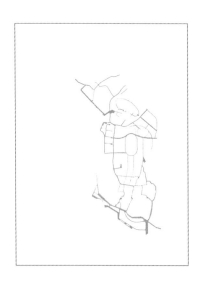

道路梳理

废旧建筑梳理

景观梳理

上篇：融合与共生 —— 华侨大学厦门校区校园空间可持续发展概念设计

"中间校区"建筑设计 I

中间校区建筑平面图

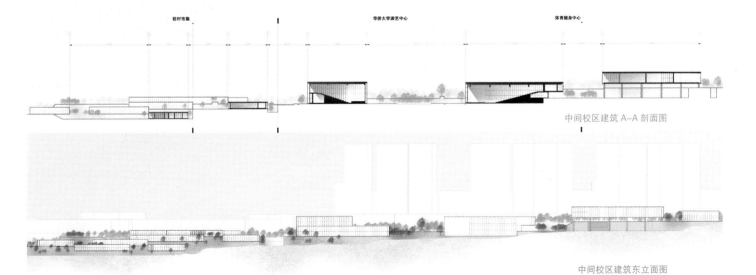

中间校区建筑 A-A 剖面图

中间校区建筑东立面图

融合·共生与介入·整合 —— 华侨大学厦门校区与村落的城市设计研究

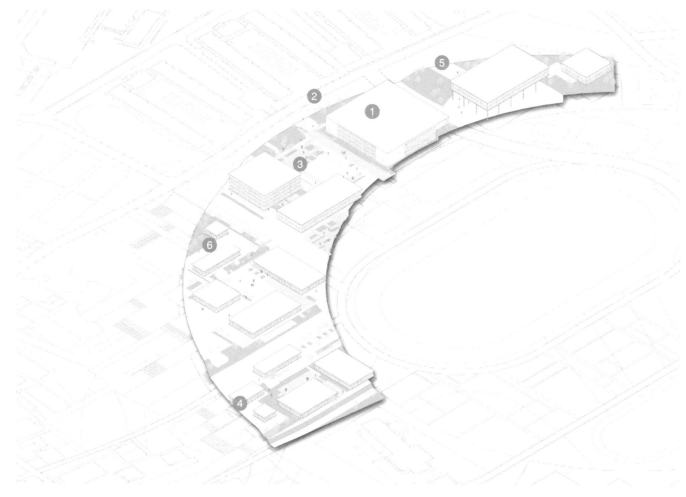

"兑山书院" 建筑设计 |

兑山书院建筑平面图

a	b	c	d
空间连接	空间围合	空间序列	融入自然

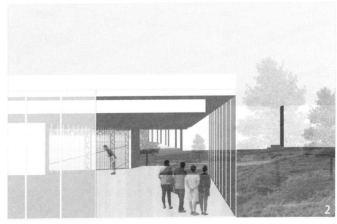

融合·共生与介入·整合——华侨大学厦门校区与村落的城市设计研究

"新乐园"环境设计 |

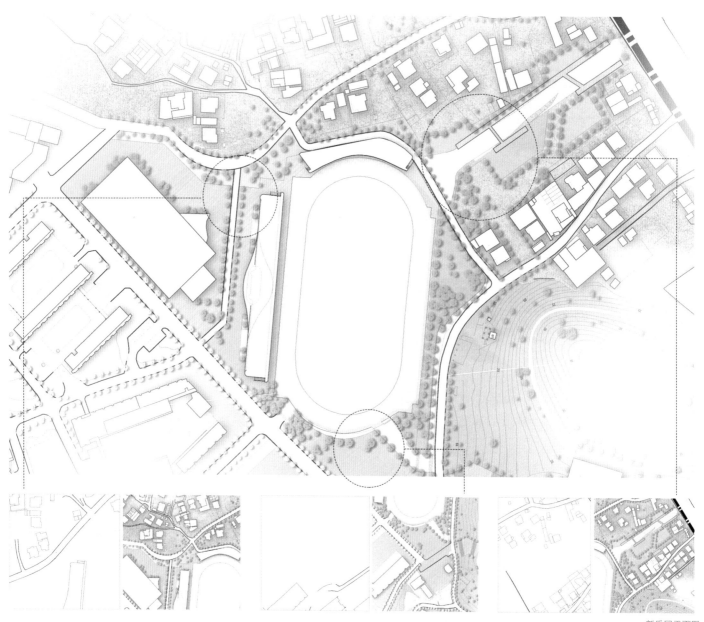

新乐园平面图

上篇：融合与共生 —— 华侨大学厦门校区校园空间可持续发展概念设计

从串连到串联

"From Connection to Communication"

"融合与共生"华侨大学厦门校区校园空间可持续发展概念设计

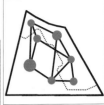

FROM CONNECTION TO COMMUNICATION

从串连到串联

指导老师：许懋彦　课程助教：弋念祖
设计者：刘玉超　徐小辉　于博

一、前期分析

应对策略

根据以上校村融合存在的问题，我们从校村双方的需求与价值角度出发。希望能够在双方的需求与价值之间建立连接的纽带。

串连——物质空间的连接
串联——精神文化的联系

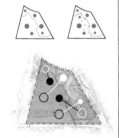

○ 需求点　● 价值点

1.文化创意产业基地　11.体育场
2.学生村民活动中心　12.文创商业区
3.榕树公园　13.会堂观演中心
4.校史档案馆　14.古榕公园
5.学生休闲商业区　15.华侨博物馆
6.大学生活动中心　16.民俗商业街
7.乒羽运动馆　17.专家公寓
8.学生宿舍区　18.体育馆
9.学生运动场　19.游泳馆
10.村民运动场　20.龙舟馆
　　　　　　　21.接待中心

校园总平面图

校园商业中心　学生休闲广场　学生街　民俗文化街
复建宗祠　校园休憩点　家庙　宗祠
运动场地　生态水系　古榕公园　古树　休闲广场

道路系统分析

教师评语：

在城市脉络发展的此消彼长中，我们应该在承认城市"历时性"的同时，也要尊重"共时性"的可能。校园与村落虽然是两种城市语境，但是也存在对话的可能性。旧有的村落具有厚重的历史习俗，人文环境。新有的校园具有良好环境。作品通过商业轴线、民俗休闲轴线和景观文化轴线的校村统筹规划，实现校与村的物质空间的新串连与旧的精神空间串联。

功能分区分析　道路系统分析

绿化景观分析　规划片区结构分析

融合·共生与介入·整合 —— 华侨大学厦门校区与村落的城市设计研究

从串连到串联 "From Connection to Communication"
"融合与共生"华侨大学厦门校区校园空间可持续发展概念设计

二、西珩南尾井片区设计

片区位置

西珩、南尾井片区位于整个校园片区的北端，整个片区传统建筑和肌理保存完整，位于片区中心。整个片区建筑密度较大，缺乏公共交流空间，景观资源丰富，但缺乏整合优化。

现场照片

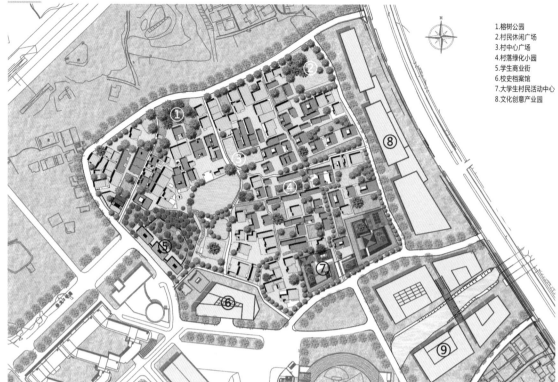

1. 榕树公园
2. 村民休闲广场
3. 村中心广场
4. 村落绿化小园
5. 学生商业街
6. 校史档案馆
7. 大学生村民活动中心
8. 文化创意产业园

西珩、南尾井片区总平面图

空间策略

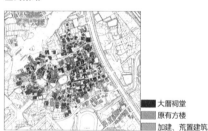

大厝祠堂　原有方楼　加建、荒置建筑

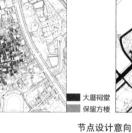

大厝祠堂　保留方楼

节点设计意向

设计说明

对于西珩、南尾井片区位采取传统建筑风貌肌理保护的策略、局部拆除的整治策略。对整个片区的公共服务设施、景观环境以及产业业态进行整合优化，植入部分校园功能，促进校村融合、共生。

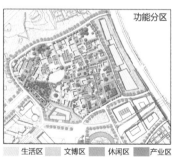

功能分区　　生活区　文博区　休闲区　产业区

道路系统　　片区主干道　片区次干道

绿化景观分析　　古树绿化节点　水系景观节点

节点分布　　广场　公园　景观绿地　文化活动

绿化节点　　　公共节点　　　校村节点

上篇：融合与共生——华侨大学厦门校区校园空间可持续发展概念设计

从串连到串联 "From Connection to Communication"
"融合与共生"华侨大学厦门校区校园空间可持续发展概念设计

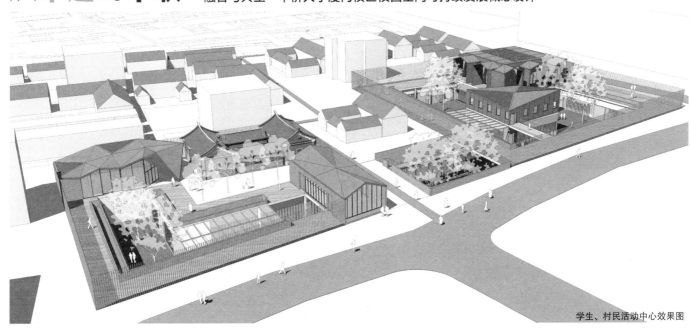

学生、村民活动中心效果图

学生、村民活动中心位于南尾井片区，处于校园村落边界处。学生社团活动中心和村民活动中心分别位于地块两侧，隔路而望，分别满足学生和村民的业余娱乐生活需要。希望将其打造成为促进学生和村民、校园和村落交流的场所。两个活动中心在地下相连通，在空间上促进了双方的交流。实现了从物质空间的串连到精神交流串联的目的。

大学生社团活动中心
1.多功能活动厅
2.排练室
3.社团活动室
4.室外活动场地
活动
5.商店
6.超市
7.咖啡厅
休闲
8.办公室
9.卫生间、设备用房
办公

村民活动中心
1.多功能活动厅
2.村民阅览室
3.村民活动室（书画、棋牌、茶道、音乐、戏曲、民俗）
活动
5.茶室
6.商店
休闲
7.办公室
8.村民会议室
9.卫生间、设备用房
办公

一层平面图

地块分离 ⇄ 通过地下空间进行贯通
串联

地下一层平面图

节点透视一

节点透视三

节点透视二

剖透视

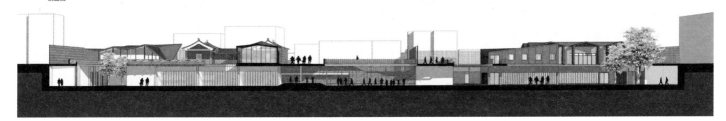

融合·共生与介入·整合 —— 华侨大学厦门校区与村落的城市设计研究

从串连到串联 "From Connection to Communication"
"融合与共生"华侨大学厦门校区校园空间可持续发展概念设计

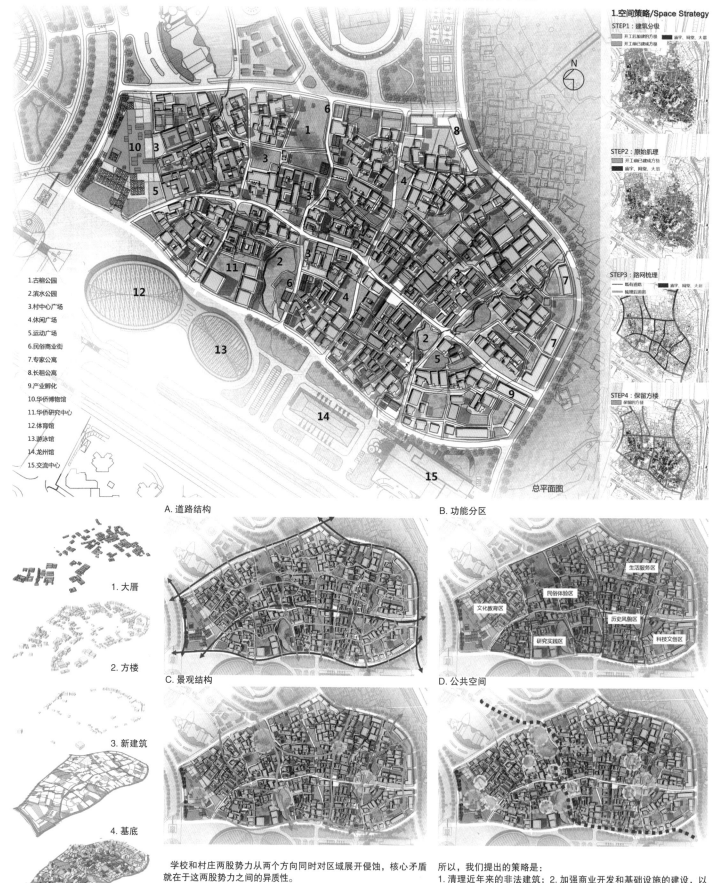

学校和村庄两股势力从两个方向同时对区域展开侵蚀，核心矛盾就在于这两股势力之间的异质性。

尤其是在中国快速城市化的背景下，城中村由于区位优势爆发式增长，成为社会中低端人群在城市的落脚点，矛盾因人而生。

所以，我们提出的策略是：
1. 清理近年来的非法建筑；2. 加强商业开发和基础设施的建设，以此推进拆改；3. 在红线范围附近，建立综合型社区，隔绝外界的干扰；
4. 逐步实现内部的升级，达成学校对区域的保护和升级；5. 华侨大学与华侨村落和谐共生。

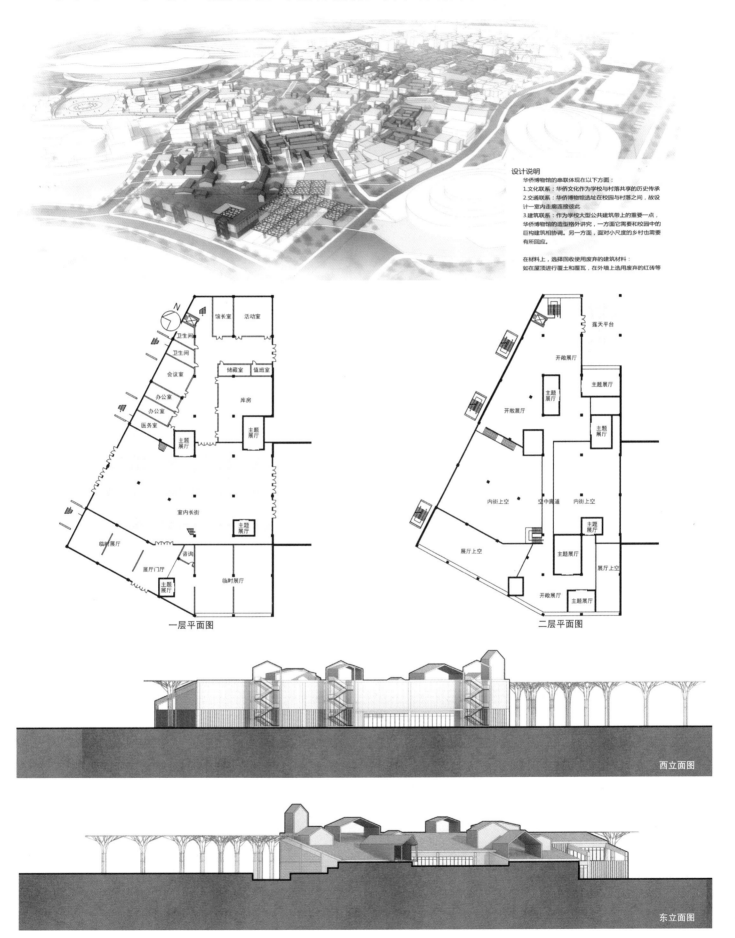

融合·共生与介入·整合 —— 华侨大学厦门校区与村落的城市设计研究

从串连到串联

"From Connection to Communication"
"融合与共生" 华侨大学厦门校区校园空间可持续发展概念设计

设计说明

通过调研分析，对中心区的更新改造策略不同于两侧南芋井和下蔡村落的改造。由于该区域大片为闲置绿地，少部分集中建楼群和商业带，相对的历史古建筑保护压力小。我们的策略是进行适度的拆建更新，与校园新区连成一片。在该区域重点植入和增补文体、休闲、商业和文创产业的功能，目的是同时吸引校区和村落的人群积极参与进来，串联校区与村落的文体与商业活动，真正实现校村的融合共生。

功能分区

公共空间

空间结构

道路系统

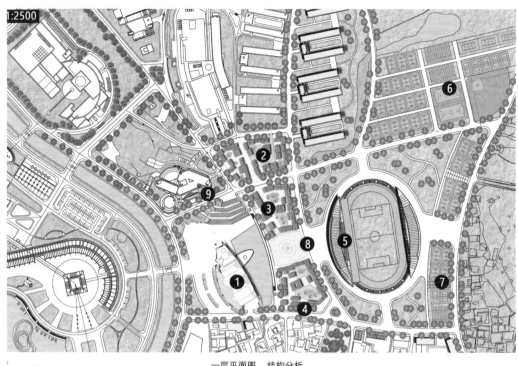

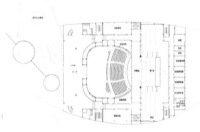

一层平面图　结构分析

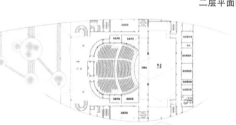

二层平面图

环境分析

会堂北侧利用场地原有高差形成室外看台，与湖面伸出台面共同构成室外剧场，可供村民及学生共同使用。

上篇：融合与共生 —— 华侨大学厦门校区校园空间可持续发展概念设计

项目概况
Introduction

回顾集美区兑山村的历史，眼下的村校矛盾只是这个聚落发展的一个中间时态。未来这里将是一个校园，但是一个校园的未来可以是怎么样的？是否有可能利用村落和侨文化以及现有自然资源，来打造一个别具一格的学校，让每个从华侨大学出来的人都染上独特的文化印记？

本设计根据校园现状将校园分成四个区域：介入微设计的现有校园部分、体验传统文化和自然生活的百草园、担当现代教学职能的埕场与以传统中国书院为原型的兑山书院。这四个区域在职能上互相补充，而在性格上各具特点，是结合了场地现状与文化的新校园形态。该设计为华侨大学村校融合提供了更多的可能性。

从百草园到三味书屋
FROM BAICAO GARDEN TO SANWEI STUDY

指导老师：许懋彦　课程助教：弋念祖
设计者：徐晨宇　卢清新　郭婧舒

四个模块
Four Modes

 学院/Academy
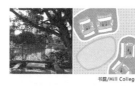 书院/Hill College
 埕场/Plaza
 百草园/Garden

校、村历史
History of School and Village

 (1)嘉山翁离，修建祖庙 / Arrival of the Ancestor
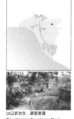 (2)以农为生，渐渐发展 / Developing by agriculture
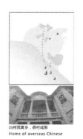 (3)村民离乡，侨村成形 / Home of overseas Chinese
 (4)水流下降，渐成滩涂 / Appearance of the shoal
 (5)集美新区，开始建设 / Construction of the new district
 (6)大学进驻，互相吴剥 / Battle between village and college
 (7)重新探讨，融合共生 / Studio on this project

校、村的愿景
Vision of School and Village

 可以是一个完全现代化的校园吗？/ Could there all be modern facilities?
 可以保存侨村的物质空间形态吗？/ Could it retain the village space form?
 可以利用良好的自然环境吗？/ Could it take use of the nature environment?
 可以继承宝贵的传统侨文化吗？/ Could it inherit the traditional culture?

总体结构分析
Structure Analysis

教师评语：

有"村"在校也许常见，有"侨"在校也可能不是唯一。两者兼备或许是华侨大学独有的特点。校园里有传统龙舟的水系，有乡野之中的百草园，有传统院落的书院，还有教学育人的埕场。村落的别致景色和侨眷的文化涵养都融汇在校园之中。未来的发展作品用微更新的手法，将未来校园发展的需求纳入村落的空间、建筑和环境里，探讨了一种尊重旧有肌理，发掘历史文脉的校园发展之道。

融合·共生与介入·整合 —— 华侨大学厦门校区与村落的城市设计研究

总平面图 Master Plan

总平面图
Master Plan

上篇：融合与共生 —— 华侨大学厦门校区校园空间可持续发展概念设计

设计方案 1：埕场
Design Project 1: Plaza

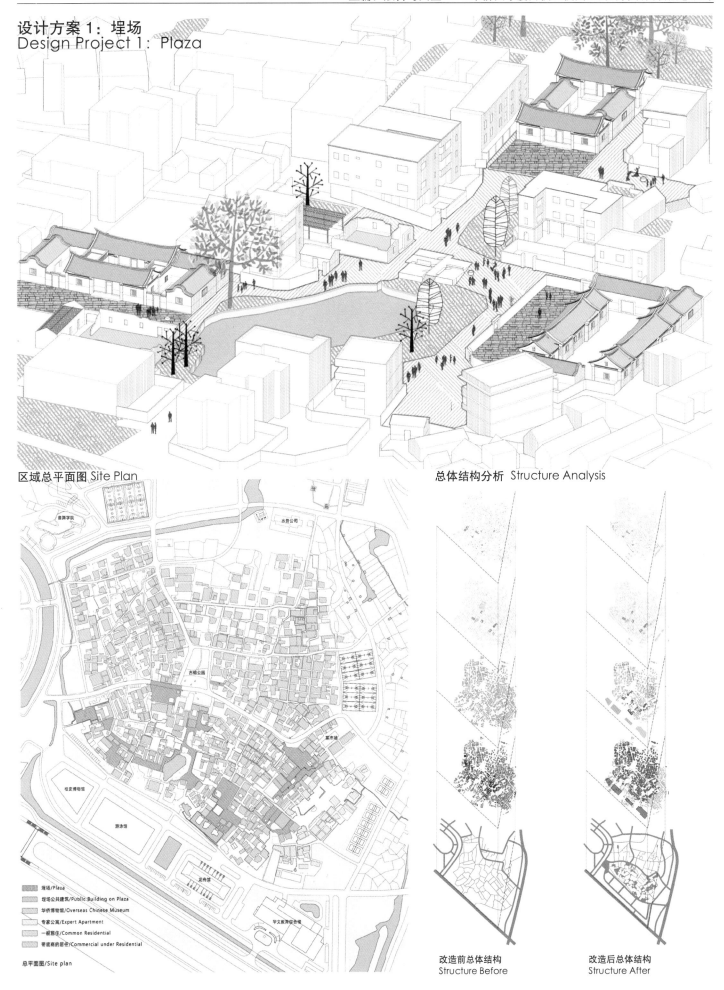

区域总平面图 Site Plan

总体结构分析 Structure Analysis

埕场/Plaza
埕场公共建筑/Public Building on Plaza
华侨博物馆/Overseas Chinese Museum
专家公寓/Expert Apartment
一般居住/Common Residential
带底商的居住/Commercial under Residential

总平面图/Site plan

改造前总体结构
Structure Before

改造后总体结构
Structure After

节点 A：侨之埕场 Node A: Plaza with Museum

首层平面图 Ground Floor Plan

功能分区 Function Zone

院落组织 Courtyard Organisation

节点 B：池之埕场 Node B: Plaza with Pond

首层平面图 Ground Floor Plan

空间轮廓 Contour of the Plaza

界面形态 Facade of the Interface

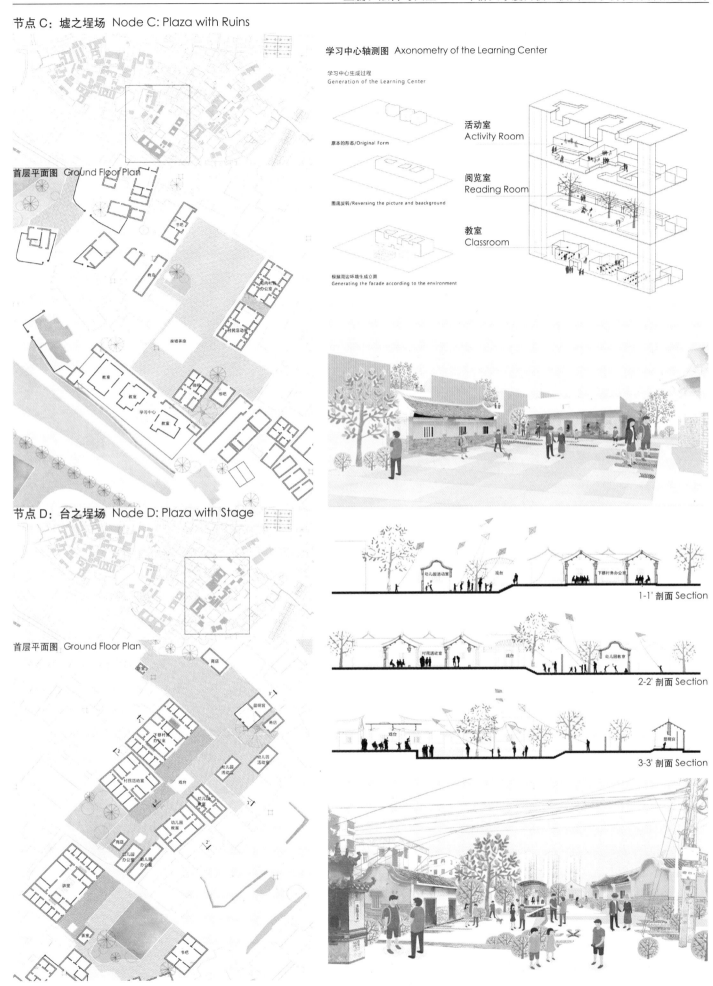

融合・共生与介入・整合 —— 华侨大学厦门校区与村落的城市设计研究

设计方案 2：兑山书院
Design Project 2: The Hill College

节点 A：遗迹公园
Node A: Ruins Park

区域总平面图 Site Plan

总体结构分析 Structure Analysis

自然与嵌套
Nature and Nesting Environment

中心与仪式
Center and Ritual Place

老厝与新书院
The Old and New College

弱势与意外
The Blurring and Randomicity

节点 B：书院综合体
Node B: College Complex

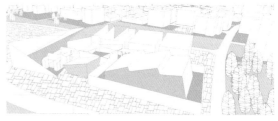

节点 C：书院
Node C: College

首层平面图 Ground Floor Plan

轴测图 Axonometric

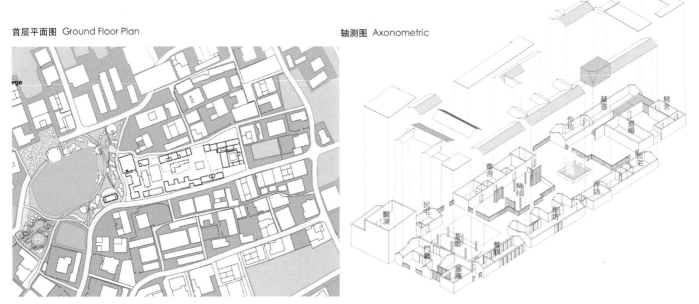

设计方案3：百草园
Design Project 3: Baicao Garden

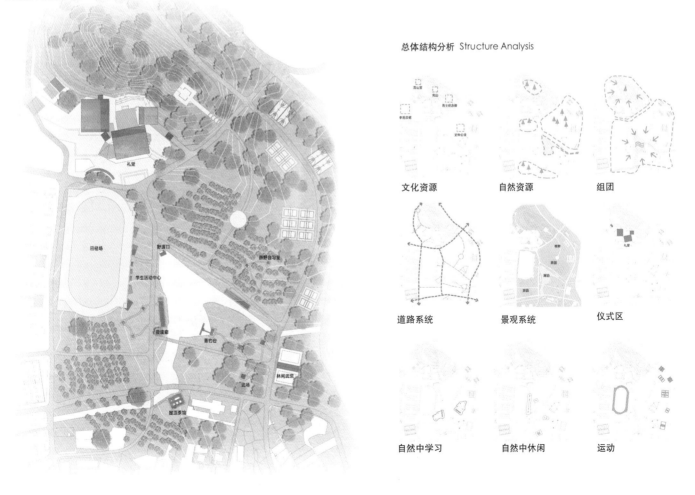

区域总平面图 Site Plan

总体结构分析 Structure Analysis

文化资源　　自然资源　　组团

道路系统　　景观系统　　仪式区

自然中学习　　自然中休闲　　运动

融合·共生与介入·整合 —— 华侨大学厦门校区与村落的城市设计研究

节点 A：礼堂
Node A: Great Hall

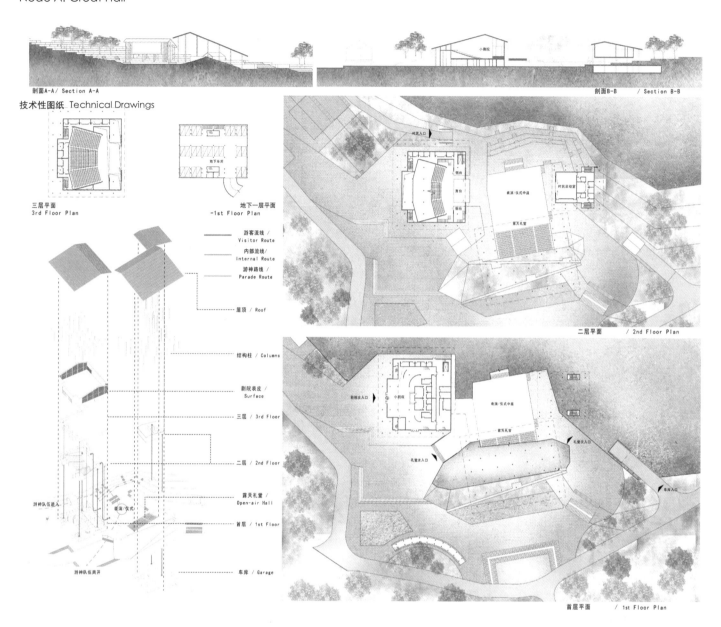

节点 B：自然中的生活
Node B: Life in Nature

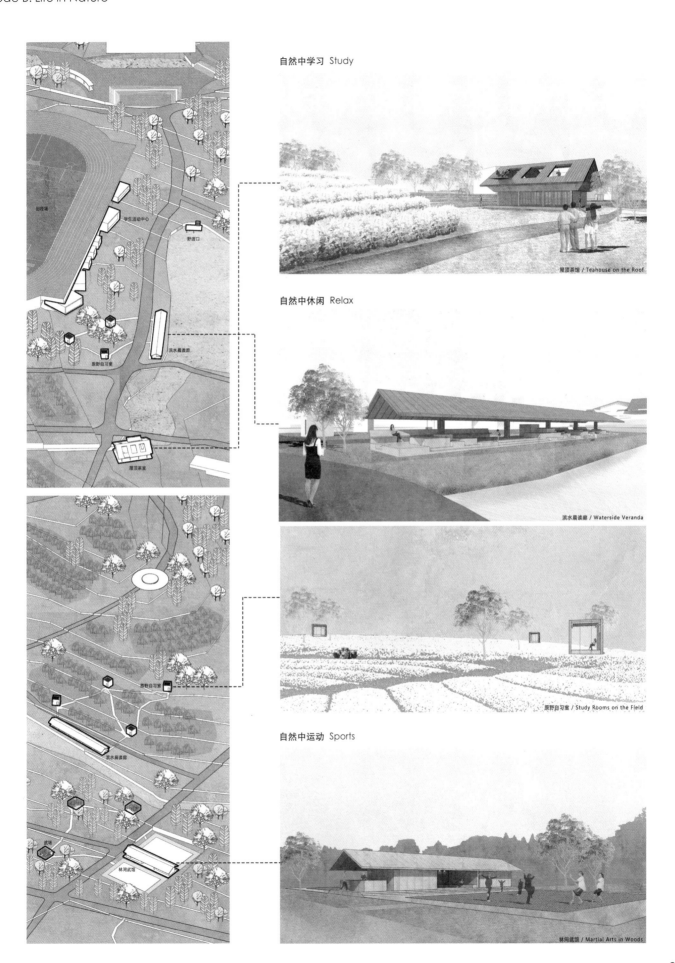

节点 C：现代化学院
Node C: Academy

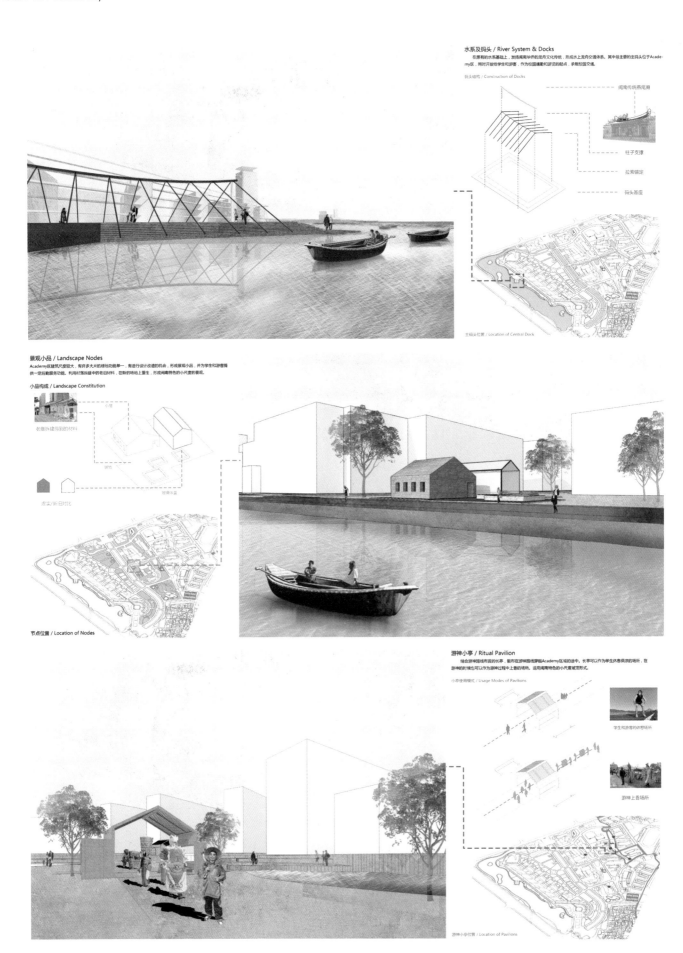

上篇：融合与共生——华侨大学厦门校区校园空间可持续发展概念设计

Available space latent in the village
Utilization of space resources of village and transfer of university function
Spatial Sustainable Development of HQU's Xiamen Campus Project

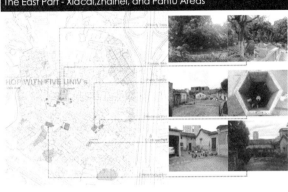

指导老师：古谷诚章　课程助教：张卓群
设计者：Hakjae Kim　陈麟　陈路　御供崇尚　仲西将　宫岛春风

AVAILABLE SPACE LATENT IN THE VILLAGE

The Space with Potential

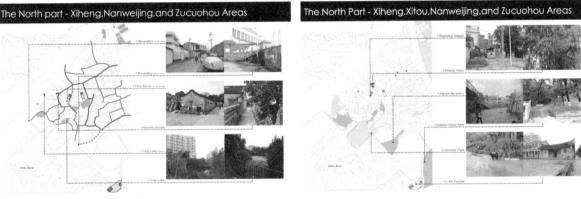

Wall　　Vacant House　　Historical Building　　Park

The Space that Began to be Utilized

Gallery　　Rest Area　　Barn　　Vegetable Garden

教师评语：

该方案首先基于详实的调查，归结出周边居住区之水色多种又水之特点，这一特点通过整个校区周围多个节点以水色之大小上和空间呈现感和质感。之后系统地尝试了周边路径及微观的方法有效地开始被广泛关注的乡村活化的项目中，细的眼光之大可以经合当地文脉和语言的话，将会是更有层次和遗憾之处是，设计更多考察和华侨大学这一特点通过整个住区周围多个节点之联系性。和处微观的方法的有效的开始被广泛关注的乡村活化项目中，细的眼光可以经合当地文脉和语言的话，将会更有层次和质感的方案。

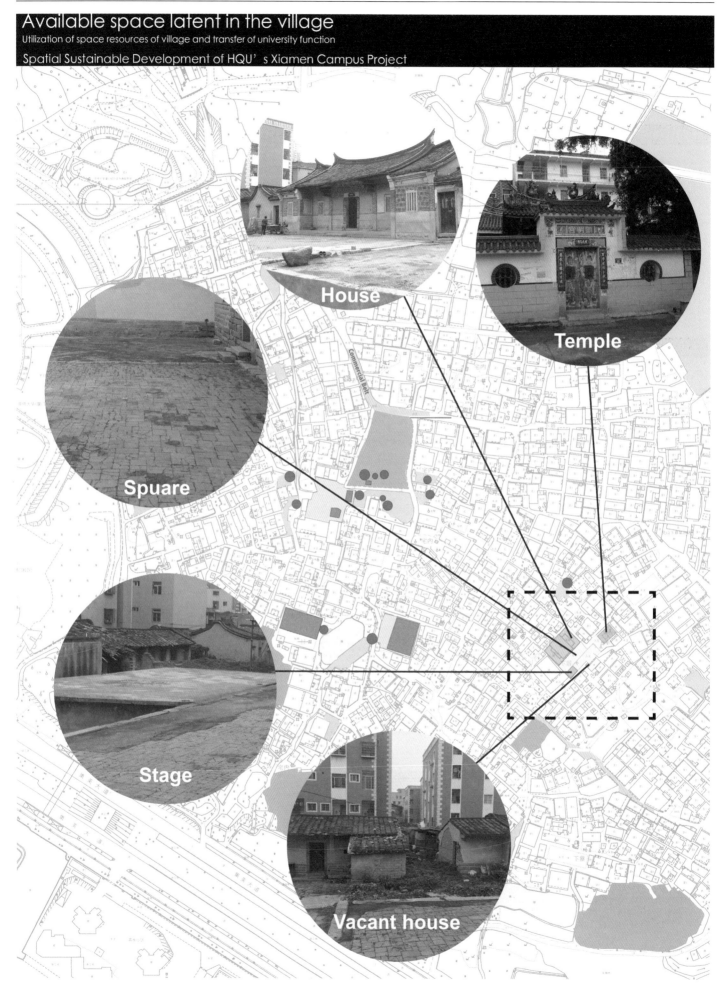

Available space latent in the village
Utilization of space resources of village and transfer of university function
Spatial Sustainable Development of HQU's Xiamen Campus Project

Proposal of Filtered Water

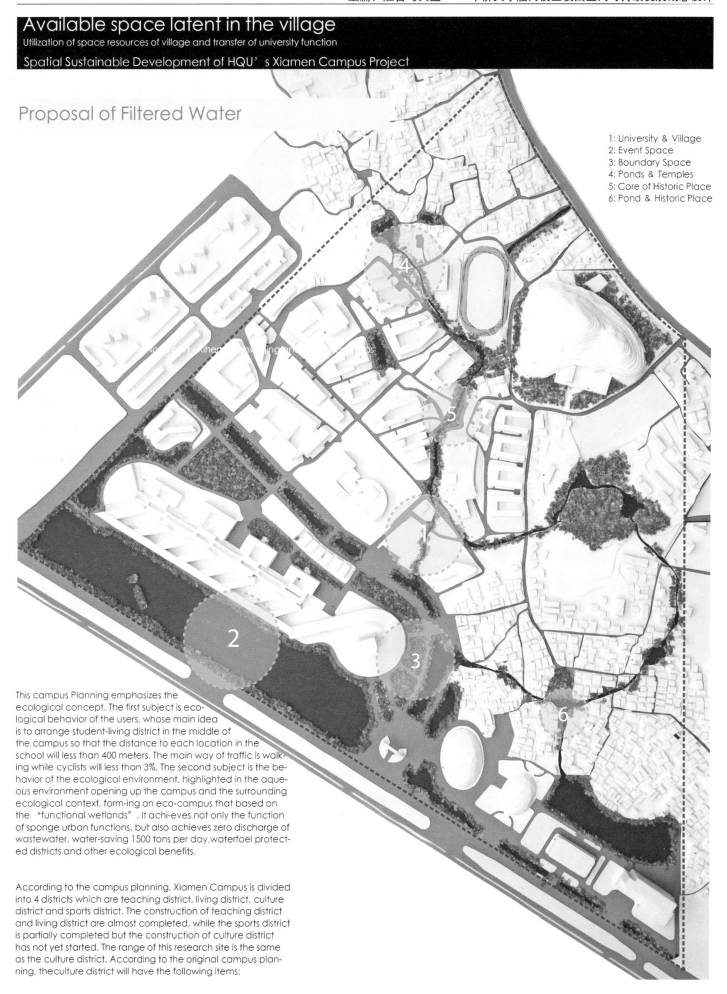

1: University & Village
2: Event Space
3: Boundary Space
4: Ponds & Temples
5: Core of Historic Place
6: Pond & Historic Place

This campus Planning emphasizes the ecological concept. The first subject is ecological behavior of the users, whose main idea is to arrange student-living district in the middle of the campus so that the distance to each location in the school will less than 400 meters. The main way of traffic is walking while cyclists will less than 3%. The second subject is the behavior of the ecological environment, highlighted in the aqueous environment opening up the campus and the surrounding ecological context, form-ing an eco-campus that based on the "functional wetlands". It achi-eves not only the function of sponge urban functions, but also achieves zero discharge of wastewater, water-saving 1500 tons per day, waterfoel protected districts and other ecological benefits.

According to the campus planning, Xiamen Campus is divided into 4 districts which are teaching district, living district, culture district and sports district. The construction of teaching district and living district are almost completed, while the sports district is partially completed but the construction of culture district has not yet started. The range of this research site is the same as the culture district. According to the original campus planning, theculture district will have the following items:

Available space latent in the village
Utilization of space resources of village and transfer of university function
Spatial Sustainable Development of HQU's Xiamen Campus Project

Green and Public space is for student because of Boundary

Once again, we expose the river and creat a green belt around. A place where villagers can be utilized not only students.

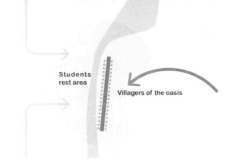

At the time of the festival set up a long table, so a lot of people get together. Become a student of the rest area, also, the villagers of the oasis. The river become a role to connect rather than walls.

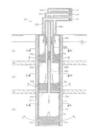

Layering process applied to wells

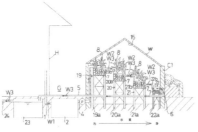

Comprehensive process applied to ponds

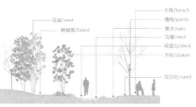

Plants purifying process applied to the center

Event Space

University & Village

Boundary Space

Core of Historic Place

Ponds & Temples

Pond & Historic Place

目录 CATALOG

引侨衔村　陈启泉　曾橙　朱宇诗
Lead Chinese Bit Village

水道侨呈　王衡　刘燚　叶桢贞
Overseas Chinese Culture Flow With Water

侨守以泮　张冬辉　任欢　陈舒舍
Keep Overseas Chinese Culture

侨脉环生　郑苍民　王志飞　刘天琪
Loop Contact

学生团队 Student Team

曾橙

陈启泉

陈舒含

刘天琪

刘燊

任欢

王衡

王志飞

叶祯贞

张冬晖

郑苍民

朱宇诗

上篇：融合与共生 —— 华侨大学厦门校区校园空间可持续发展概念设计

壹 侨脉环生
华侨大学厦门校区 "融合与共生" 概念设计 LOOp COntact

| 整 | 体 | 鸟 | 瞰 | 1
Conceptual Design of the Xiamen Campus

侨脉环生 LOOP CONTACT

指导老师：邓蜀阳　阎波　翁季
设计者：郑苍民　王志飞　刘天琪

表现图 /Perspective

设计说明：
校园规划设计以"侨脉环生"为主题，旨在体现华侨文脉，以"环"的设计手法来弥补城中村与华侨大学校园之间的裂痕，通过在场地原有文化建筑和重要节点的基础之上打造校园和城中村的重要文化或者景观节点的主要环线。在主环的基础上，辅以景观、运动、文化三个次级环线，环环相扣，使得华侨文脉在场地得以体现，同时也弥补校园原有缺失功能，为华侨大学的进一步发展提供相应的物质和文化空间基础。

Design description:
Campus planning and design to overseas vein perichaetine "as the theme, to reflect the Chinese context, the rift between the villages and the Chinese University Campus with a" ring "of the design method to make up the by on the basis of the original site cultural construction and an important node of the to build campus and villages in the important culture or landscape nodes, on the main ring road. On the basis of the Lord of the rings, supplemented with landscape, sports and cultural three secondary loop, interlocking, the overseas Chinese context at the venue embodied, but also to make up for the original campus loss of function, provide the corresponding material and cultural space foundation for the further development of the overseas Chinese University.

概念提出 /Conception

脉 Overseas Chinese Context + 环生 loop contact

概念演绎 /Inference

step1.抽象血管动脉的结构　step2.毛细血管渗入校园及城中村　step3.活力因子在动脉中循环

功能融合 /Functional Fusions

空间结构分析 /Space Structure of the Village

场地分析 /Location Analysis

学校与村落范围线 Land Area　土地建设情况 Land Development　校村路网分析 Traffic Actuality　校村绿化水体分析 Green Water
校村公共空间分析 Public Space　公共交通分析 Public Transport　周边功能分析 Peripheral Services　校园建筑分析 Campus Architecture

教师评语：
作品梳理出四个关键问题，如在整体环境上如何弥补校村差异、实现在校园肌理上如何塑造校村路网结构、如何以及造校园环境、功能上如何实现校村共融、文化特色上如何打造等，并制定了相应的策略，通过环状联系侨文化脉络，实现校村之间的融合与共生。

融合·共生与介入·整合 —— 华侨大学厦门校区与村落的城市设计研究

贰 侨脉环生
华侨大学厦门校区 "融合与共生" 概念设计　Loop Contact

| 规 | 划 | 结 | 构 | 2 |

Conceptual Design of the Xiamen Campus

总平面图 /General Plan

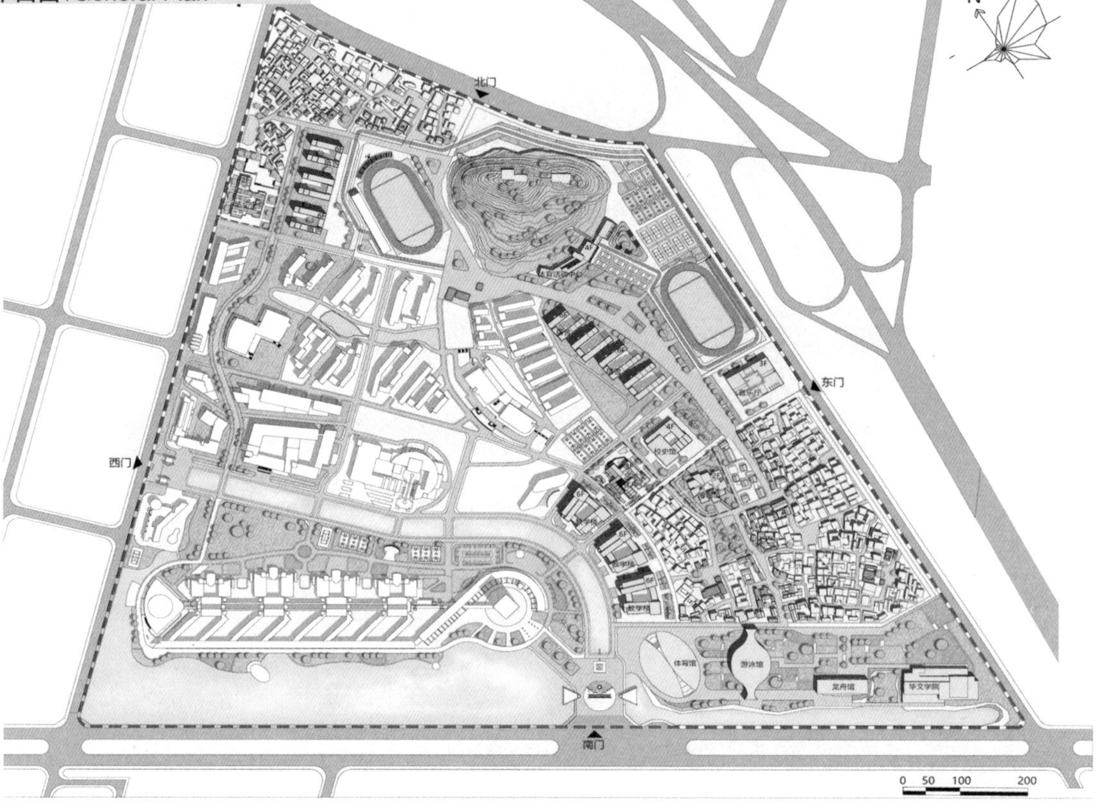

村落建筑分析 /Village Architecture Analysis

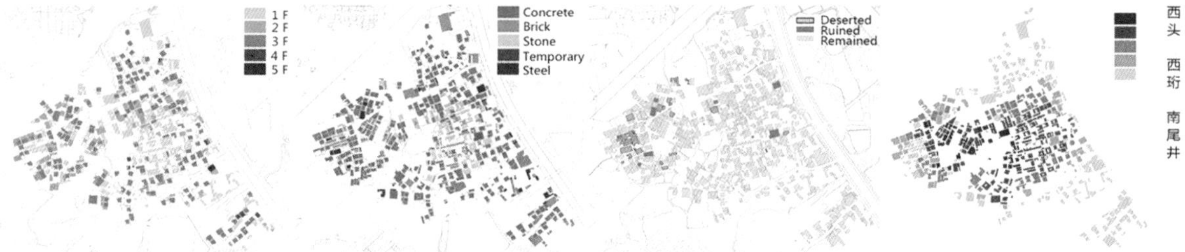

建筑高度 Height Distribution　建筑材料 Building Material　使用状况 Building Condition　建筑分级 Building Classfication

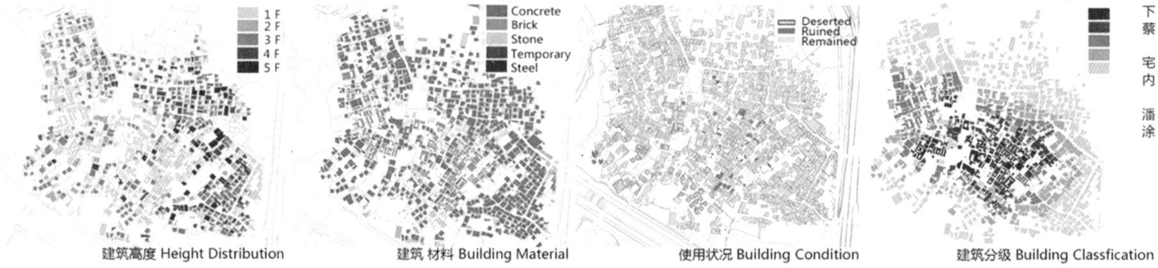

西头　西珩　南尾　井

建筑高度 Height Distribution　建筑材料 Building Material　使用状况 Building Condition　建筑分级 Building Classfication

下蔡　宅内　潘涂

规划结构分析 /Planning Structure Analysis

路网结构 Traffic Network Structure　功能分区 Functional Partitioning　环线分析 Loop Analysis　建设单体 Important Buildings

38

垣 . 生

华侨大学厦门校区 艺术文化中心建筑设计 Boundary Reborn

单 | 体 | 设 | 计 3
Art and Culture Center Design

表现图 /Perspective

经济技术指标 / Economic and Technical Indexes

总用地面积 Total land area：7615㎡
总建筑面积 Total construction area：5606㎡
容积率 Volume ratio：0.73
建筑密度 Building density：26.5%

设计说明 Design Instruction

本案位于城中村与校园交界区，是校园文化与城中村文脉交织融合的区域，所以本案提取城中村古厝坡屋顶形式，加入现代开窗及平屋顶要素，达到新旧建筑形式的融合。局部加入闽南骑楼的拱圈要素，增加灰空间活动区域。建筑功能适应周边多样化人群的使用需求，增加艺术培训及文化展览等校园缺失功能，形成一个多人群互动融合的多元化空间。

The case in villages and campus junction, campus culture and city village context at the junction of the fusion area, so the case extraction Village Gu CuO pitched roof style, adding modern windows and flat roof elements, achieve the integration of the form of new and old buildings. Arch elements locally adding its Fujian gray space, increase the activity area. Building functions to adapt to the needs of the surrounding diversified population, increase the art of training and cultural exhibition and other campus loss function, the formation of a multi population interaction and integration of diversified space.

重庆大学 刘天琪
CQU/LIU TIANQI

总平面图 /General Plan

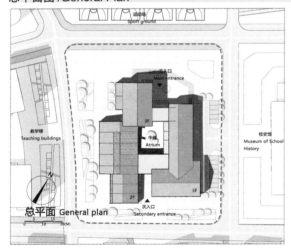

总平面 General plan

体量生成 /Volume Generation

场地位于城中村与校园的交界处
The site is located at the junction of the village and campus.

场地周围有多种使用人群
There are a variety of people around the site

确定场地入口方向
Determine the direction of the site entrance

形成街巷院落关系，聚集人流
Form the courtyard streets, gathered people

围绕院落设置交通及活动空间
Set up traffic and activity space around the courtyard

提取城中村古厝坡屋顶元素
Extract the village slope roof elements

结合周边环境动静分区
Combined with the surrounding environment, set up static and dynamic zoning

坡屋顶加入现代窗洞
Combine slope roof with modern window

建筑立面 /Facade Design

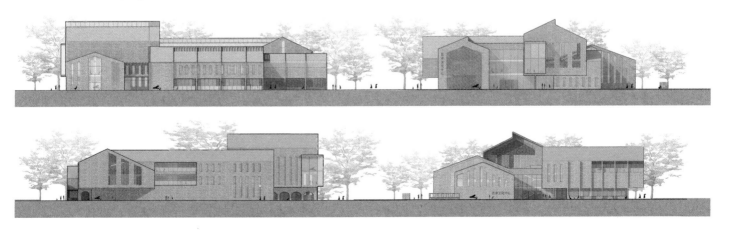

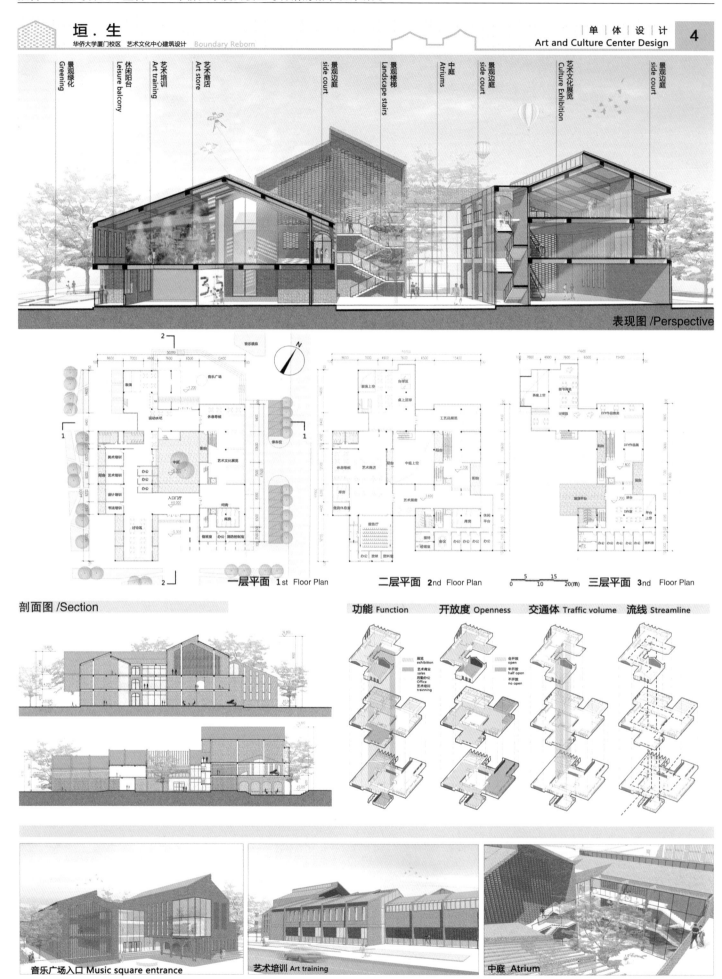

伍 方楼新生 | Reborn of the Square Buildings
华侨大学厦门校区 侨文化博物馆建筑设计
设计者：王志飞 Designer: Wang Zhifei

单体设计 5 — Museum for the Culture of overseas Chinese

效果图 Rendering

设计说明 Design Description:
We focus on the urban village which forms in the rapid process of urbanization, almost everyone of us think that the important nodes of the village should be preserved and the culture of the village should be inherited. But one important phenomenon of the urban village should not be ignored, that is the new square buildings built on the formation of urban village. Here, I choose one site which is in the north of the ancient banyan park and design one museum. My strategy is demolishing some of the square buildings and they will serve as exhibition space, then build some new blocks to combine with the square buildings. They will be important spaces to serve the campus and villagers.

总平面图 /General Plan | 体量生成 /Volume Generation

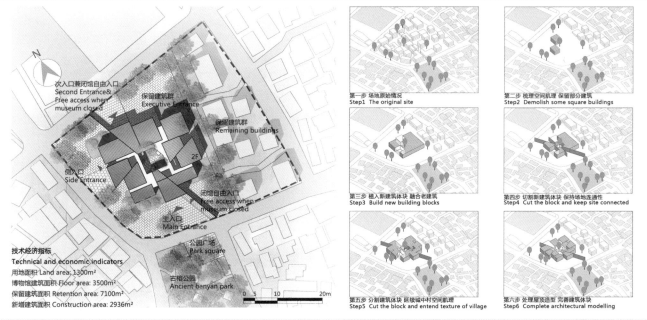

技术经济指标 Technical and economic indicators
- 用地面积 Land area: 1300m²
- 博物馆建筑面积 Floor area: 3500m²
- 保留建筑面积 Retention area: 7100m²
- 新增建筑面积 Construction area: 2936m²

第一步 场地原始情况 Step1 The original site
第二步 梳理空间肌理 保留部分建筑 Step2 Demolish some square buildings
第三步 植入新建筑体块 融合老建筑 Step3 Build new building blocks
第四步 切割新建筑体块 保持场地连通性 Step4 Cut the block and keep site connected
第五步 分割建筑体块 延续城中村空间肌理 Step5 Cut the block and entend texture of village
第六步 处理屋顶造型 完善建筑体块 Step6 Complete architectural modelling

建筑立面 /Facade Design

融合·共生与介入·整合 —— 华侨大学厦门校区与村落的城市设计研究

陆 | 方楼新生　设计者：王志飞　Designer: Wang Zhifei
华侨大学厦门校区　侨文化博物馆建筑设计　Reborn of the Square Buildings

单 | 体 | 设 | 计　Museum for the Culture of overseas Chinese　6

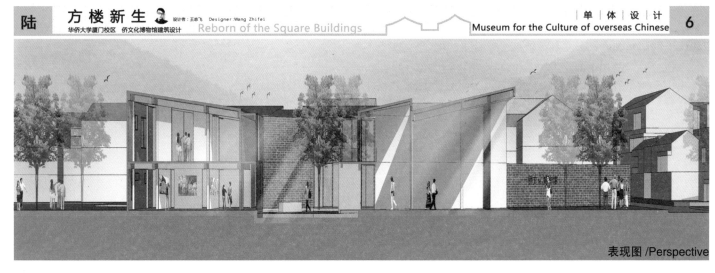

表现图 / Perspective

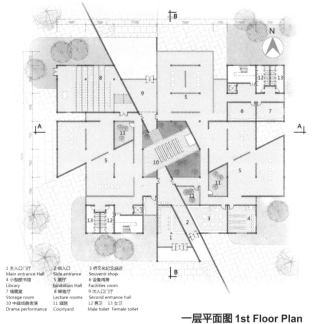

一层平面图 1st Floor Plan

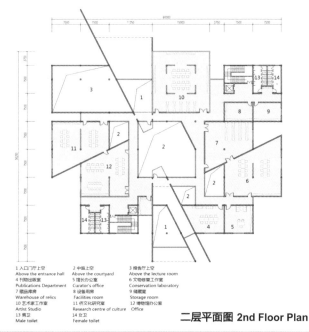

二层平面图 2nd Floor Plan

从建筑西侧看古榕公园 / West side of the Museum

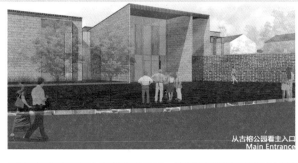

从古榕公园看主入口 / Main Entrance

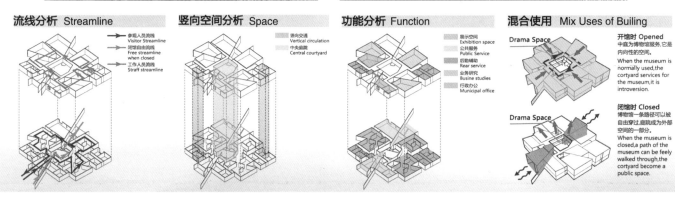

流线分析 Streamline　　竖向空间分析 Space　　功能分析 Function　　混合使用 Mix Uses of Building

上篇：融合与共生 —— 华侨大学厦门校区校园空间可持续发展概念设计

校园客厅
华侨大学厦门校区 校园活动中心
LOOp Contact

单 体 设 计

Conceptual Design of the Xiamen Campus

表现图 Perspective

总平面图 /General Plan

体量生成 /Volume Generation

提取场地周边城中村部分建筑体量和形态，由小体量建筑的组合，提取演变生成学生活动中心的形态基础

聚集　广场+绿化　体量

连接　坡屋顶　完善形态

Design description:
Construction monomer design site is located in the northwest of base, is a village in the city and campus at the intersection of design based on making up for the loss of function of campus as a starting point, build overseas Chinese university students for exhibition, academic exchanges, reading activity places. In the architectural form, select the part of the village of the city construction, restructuring and transformation of the building, the construction of the entrance square as a precursor to the campus, the city village, the city's gathering place. Inside the body, the courtyard is set up to create the activity center for the use of college students.

设计说明：
建筑单体设计场地位于基地西北处，为城中村与校园交汇之处，设计基于弥补校园功能缺失为出发点，打造华侨大学学生用于展览、学术交流、阅览的活动场所。在建筑形态上，选取城中村部分建筑形态，进行重组与改造之后生成，建筑前廊的入口广场作为校园、城中村、城市的聚集地，体量内部设置庭院丰富空间层次，以塑造出适合大学生使用的活动中心。

剖面图 /Section

建筑立面 /Facade Design

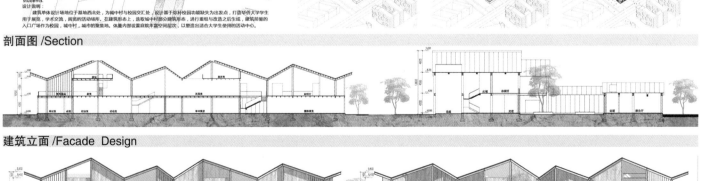

上篇：融合与共生 —— 华侨大学厦门校区校园空间可持续发展概念设计

侨守·以泮
——以文化景观资源保护为主导的校园空间可持续发展概念设计

概念策略 Concept Strategy 1

侨守以泮

KEEP OVERSEAS CHINESE CULTURE

指导老师：邓蜀阳 阎波 翁季
设计者：张冬辉 任欢 陈舒含

核心问题 Key Problem

主体	校园建设 VS 侨文化景观资源保护
诉求	华侨大学校园扩建需求、规划结构的完善；兑山村物质资源的修缮、文化资源的传承
融合	华侨大学对侨文化的传承；发展的义务与优势；侨乡建筑的价值、侨文化的德育作用

设计目标 Design Goals

整体规划视角：区域特色定位
校园规划视角：校园特色定位
侨文化保护视角：形象定位
侨文化保护视角：功能定位

策略生成 Strategy Generation

侨(qiáo) 守(shǒu) 以(yǐ) 泮(pàn)
泮：古代学宫
"守"……？文化景观

有形的、物质的	无形的、非物质的
华侨建筑 广场 古榕 古井 池塘 村落格局	民俗 节日 建筑风格 侨性的体现

Assigned to the function of the overseas chinese building campus inheriting its architectural culture.

校园策略 Campus Strategy

兑山 Dui mountain
文化广场 Culture square
主校门 Main gate

村落策略 Village Strategy

村落原始特征：
道路系统混乱；
建筑形式交杂；
建筑散乱布置；
环境恶劣
Original village feature: Road system messy; Architectural form mixed; Building scattered zero; Harsh

村落节点的保留与更新可以起到促媒作用，带动周边建筑的风貌和谐。
Village of the node can have effect on promoting media keep and update, improve surrounding buildings for harmony style.

拓宽和连续村落原有道路，形成鱼骨状道路肌理。
Broaden and confine village original road, forming herringbone road texture.

将零散建筑围合成院落形态，形成组的校园功能用房。
Scattered buildings around synthetic compound form, form groups of campus by the function.

场所保留 文脉延续

北部村落保护优先级 南部村落优先保护级
生态要素 村落空间 核心保护区

1. 传统侨乡村落外部空间要素

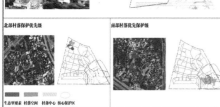
街 巷 呈 井 桥

2. 南部村落梳理后的整体形态

3. 北部村落梳理后的整体形态

厝呈
邻里呈
祠庙呈

校村策略 School Strategy

原始保留肌理 — 植入校园尺度联系 — 织补保留肌理

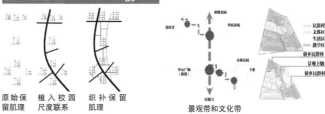

南尾 — 李氏家庙 — 民居村落区 文化区 生活区 教学区 侨乡民居 景观轴 侨乡民居村
中心广场（新建）— 古榕公园 — 下店
景观带和文化带

教师评语：

对校园建设来说，需要完善校园结构、满足师生教学空间的要求，而对侨村来说，需要对村落的物资资源进行保护修缮以及非物质文化的传承。作品敏感地认识到这两者之间的矛盾，提出"侨守以泮"这一概念，通过梳理校村整体形态、提取传统文化元素，把文化景观资源保护作为主导校园空间可持续发展的主要理念，既满足校园总体建设，又保护侨文化资源。

策略支撑

融合与共生·2016国际五校联合设计
Merging and Symbiosis · 2016 International Joint Studio
重庆大学 CHONGQING UNIVERSITY
指导教师：邓蜀阳 阎波 翁季
团队成员：张冬辉 任欢 陈舒含

融合·共生与介入·整合 —— 华侨大学厦门校区与村落的城市设计研究

侨守·以泮
——以文化景观资源保护为主导的校园空间可持续发展概念设计

规划展示 Plan Show 2

The concept design of sustainable development of campus space, which is dominated by the cultural landscape resources protection

总平面 Total Plan

设计分析 Design Analysis

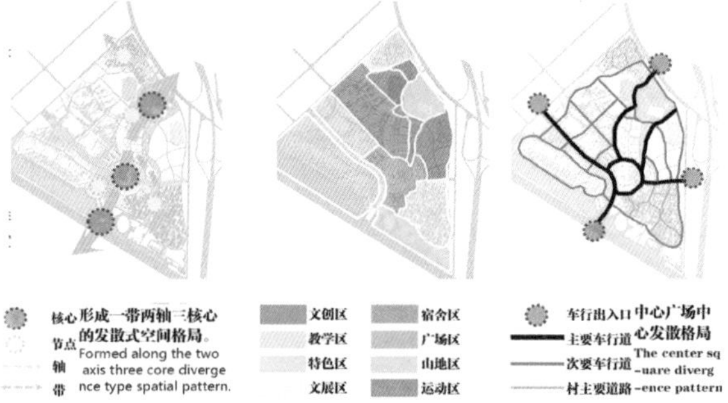

核心形成一带两轴三核心的发散式空间格局
Formed along the two axis three core divergence type spatial pattern.

- 文创区
- 教学区
- 特色区
- 文展区
- 宿舍区
- 广场区
- 山地区
- 运动区

车行出入口 中心广场中
主要车行道 心发散格局
次要车行道 The center square diverg
村主要道路 -ence pattern

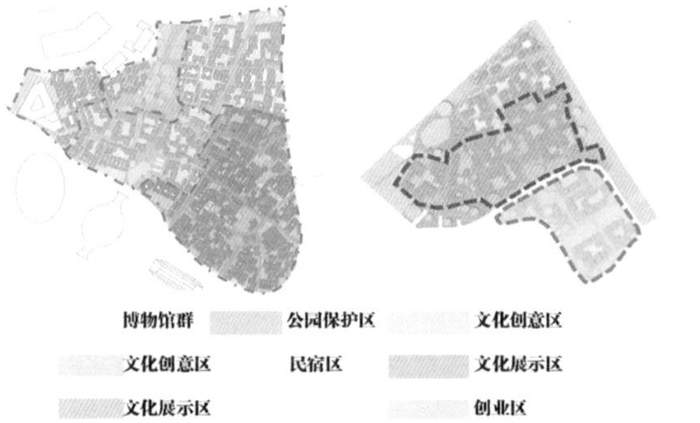

- 博物馆群
- 文化创意区
- 文化展示区
- 公园保护区
- 民宿区
- 文化创意区
- 文化展示区
- 创业区

重要地段设计 Important lot Design

文化广场

景观文化带

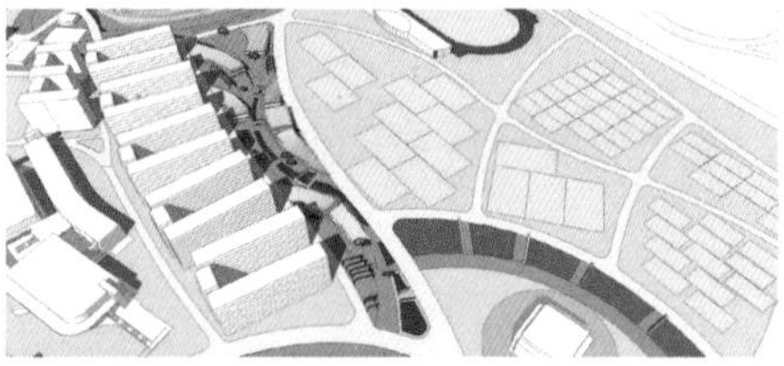

鸟瞰图 Aerial View

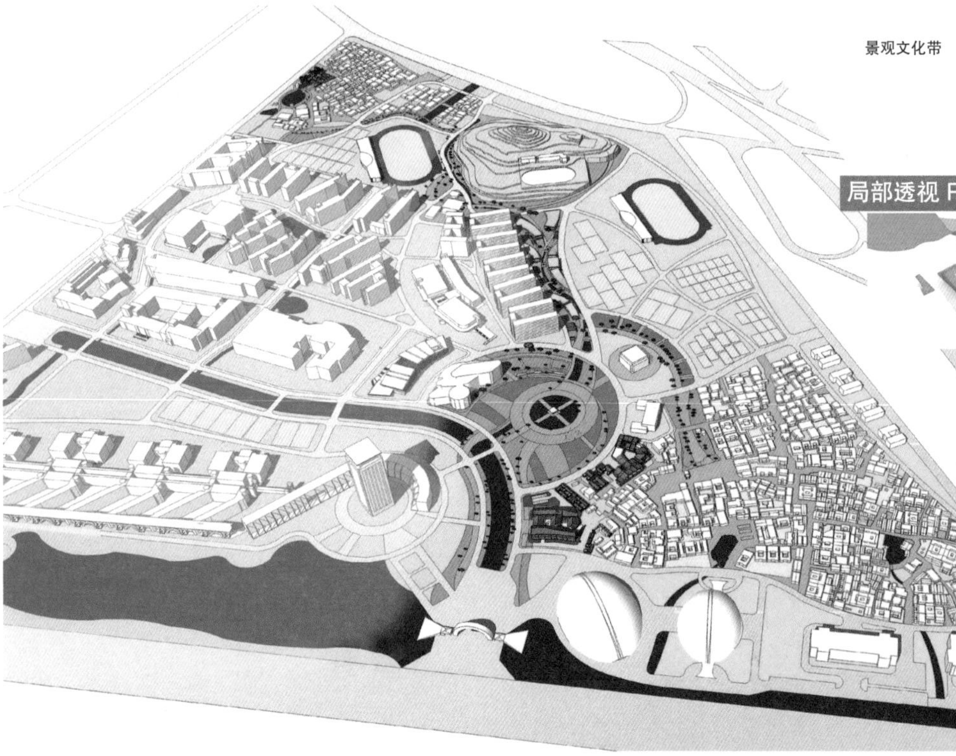

局部透视 Partial Perspective

歌仔戏研究室

大学生活动中心

华侨大学博物馆

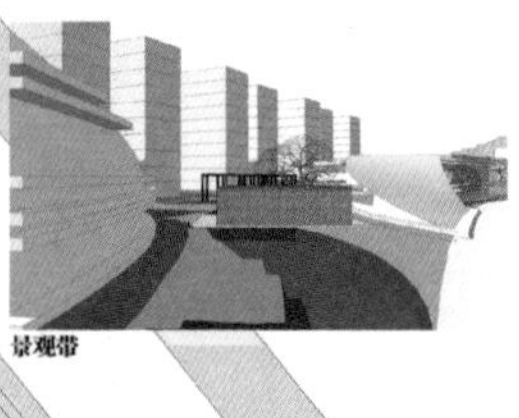

景观带

融合与共生·2016 国际五校联合设计
Merging and Symbiosis · 2016 International Joint Studio

重庆大学 CHONGQING UNIVERSITY
指导教师：邓蜀阳 阎波 翁季
团队成员：张冬晖 任欢 陈舒含

上篇：融合与共生——华侨大学厦门校区校园空间可持续发展概念设计

侨守·以泮
——以文化景观资源保护为主导的校园空间可持续发展概念设计
The concept design of sustainable development of campus space, which is dominated by the cultural landscape resources protection

大学生活动中心设计 3
University Activity Center Design

前期分析 Pre-Analysis

目标 Goal	精神场所	文化活动	文脉传承	说明 Description	方案以保留场地内4栋古厝的功能研究为主，打造侨文化活动下的大学生活动中心。The program to retain the venue of the four ancient house features mainly to create overseas Chinese cultural activities undergraduate activities center.
位置 Location	文化广场	村校边界	古建遗存	形态 Pattern	保留古厝 / 抽象古厝 / 创新古厝
功能 Features	文化研究	放映看台	文体活动	元素 Element	兑山泮池 / 大唐柿比 / 村落中心

概念构思 Concept Idea

step1 评估 assessment / step2 拆除 demolition
step3 改造 transform / step4 扩建 extension
step5 消融 ablation / step6 协调 coordination
step7 院落 yard / step8 成形 molding

总平面 Master Plan

经济技术指标：
用地面积：4615.56m²
基底面积：2478.5m²
建筑面积：3959.89m²
建筑密度：53.7%
容积率：0.86

放映区出入口
活动区出入口
研究区出入口

场地分析 Site Analysis

联系 contact
庭院 yard
空间 space
轴线 axis
流线 streamline
交流 contact

场地分析 Site Analysis

新建筑 The new building
屋面板 Roof panel
现代屋架 Modern roof truss
现代结构 Modern structure
传统结构 The traditional structure
新老地板 new and old floor

鸟瞰图 Aerial View

主要透视 The Main Perspective

侨·望——大学生活动中心设计1

融合与共生·2016 国际五校联合设计
Merging and Symbiosis · 2016 International Joint Studio

重庆大学 CHONGQING UNIVERSITY
指导教师：邓蜀阳 阎波 翁季
团队成员：张冬晖 任欢 陈舒合

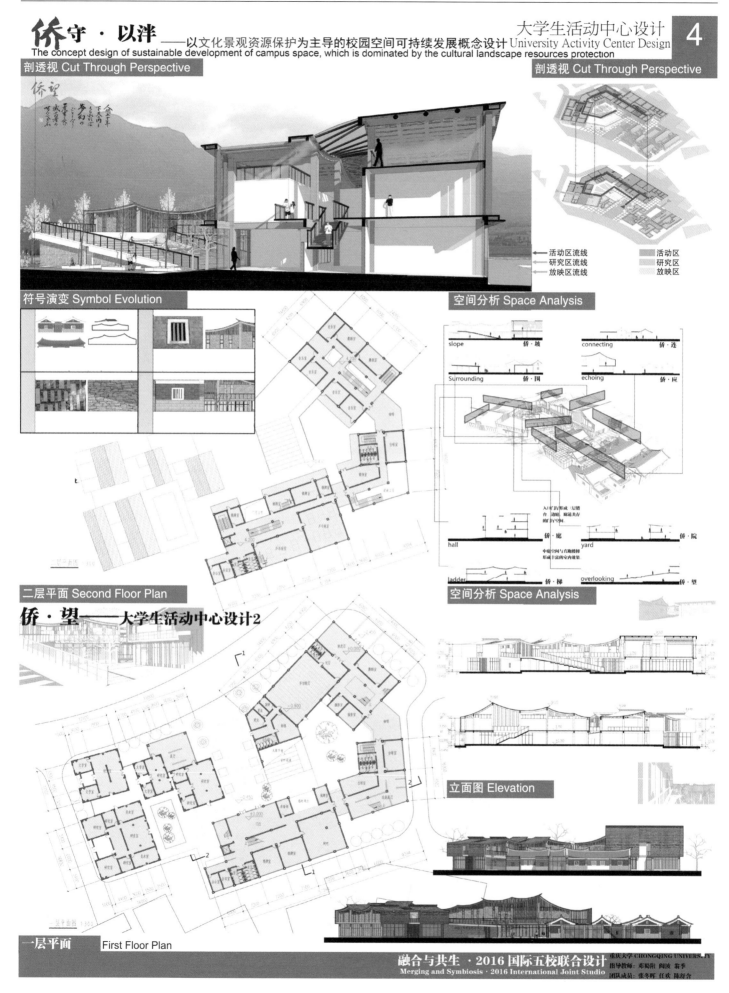

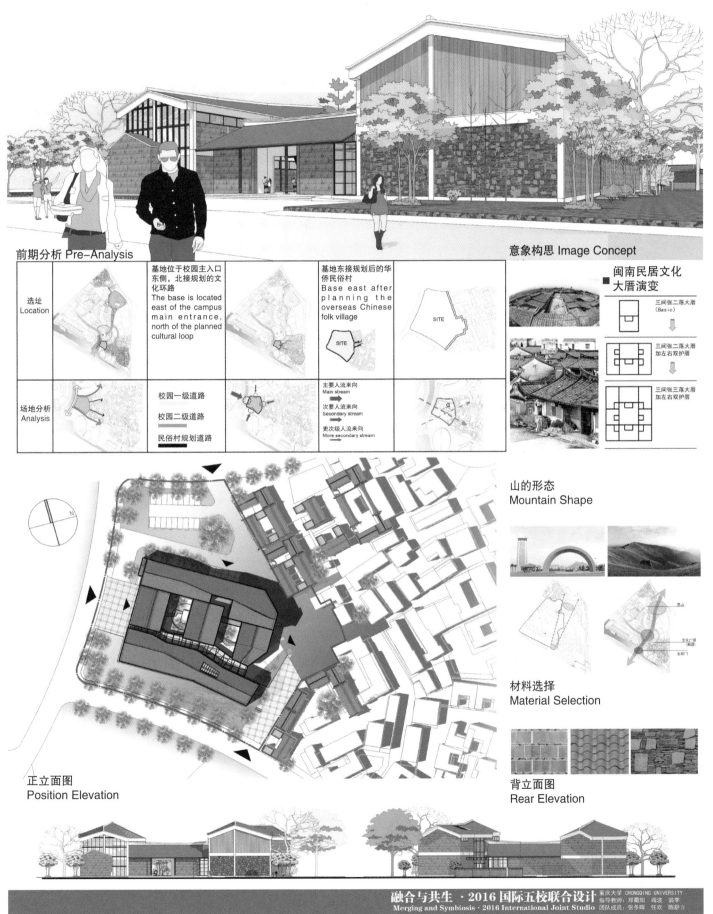

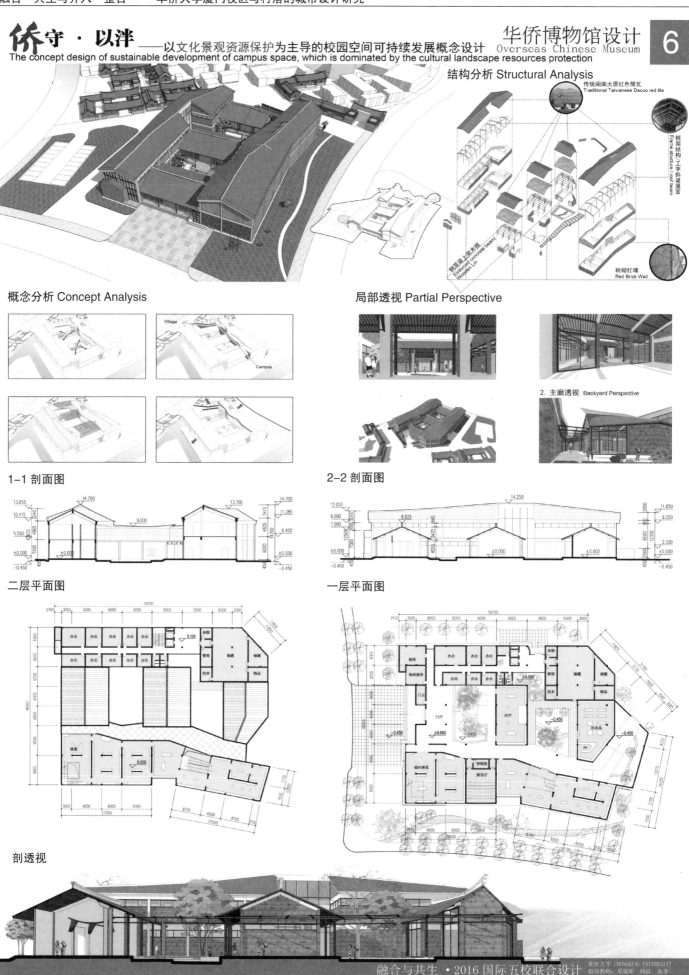

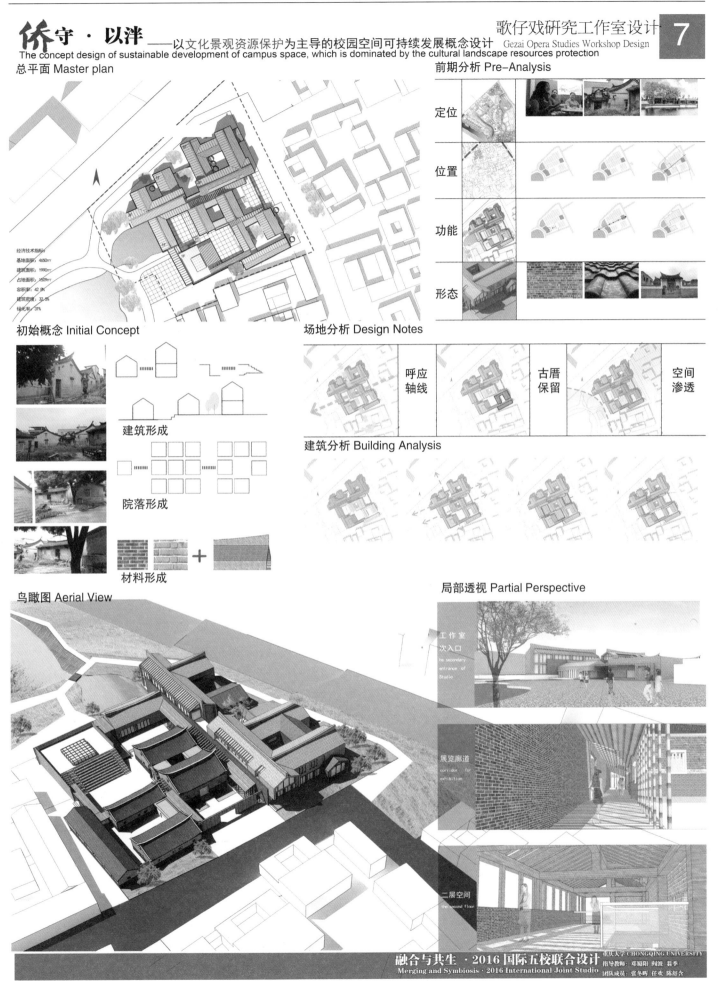

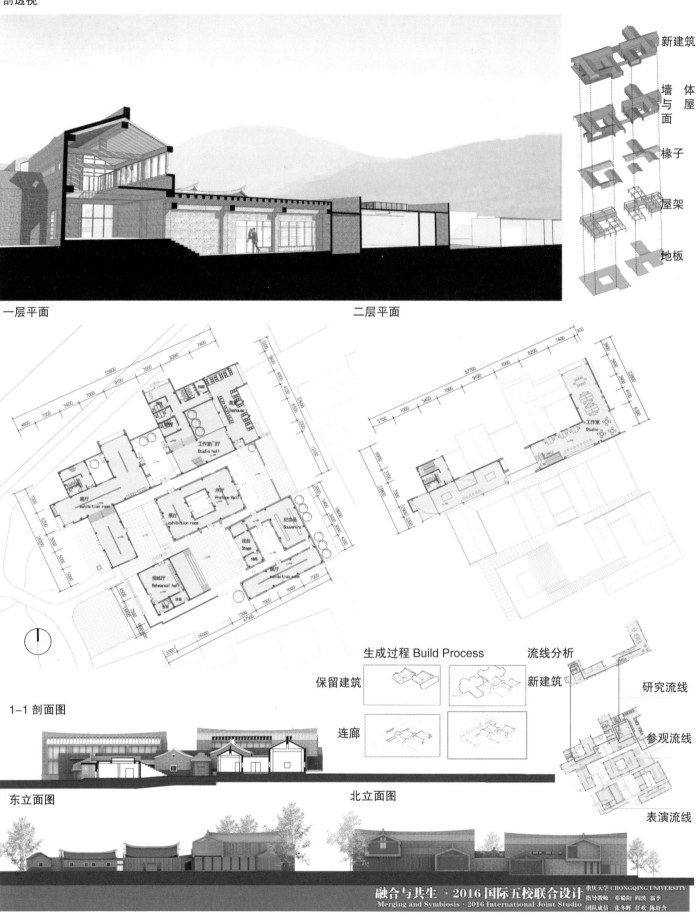

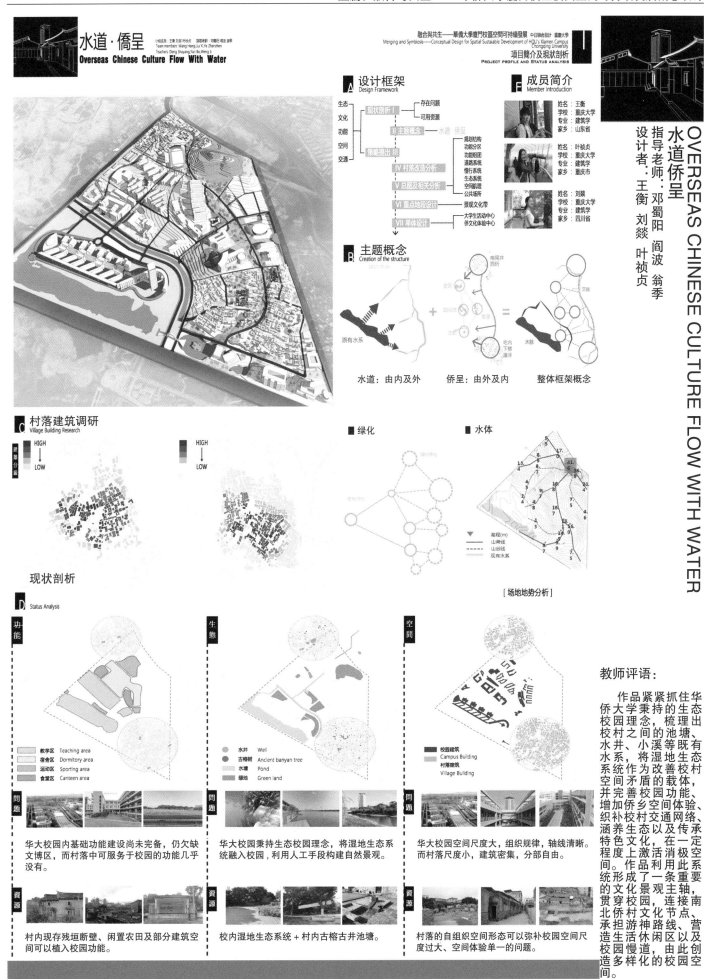

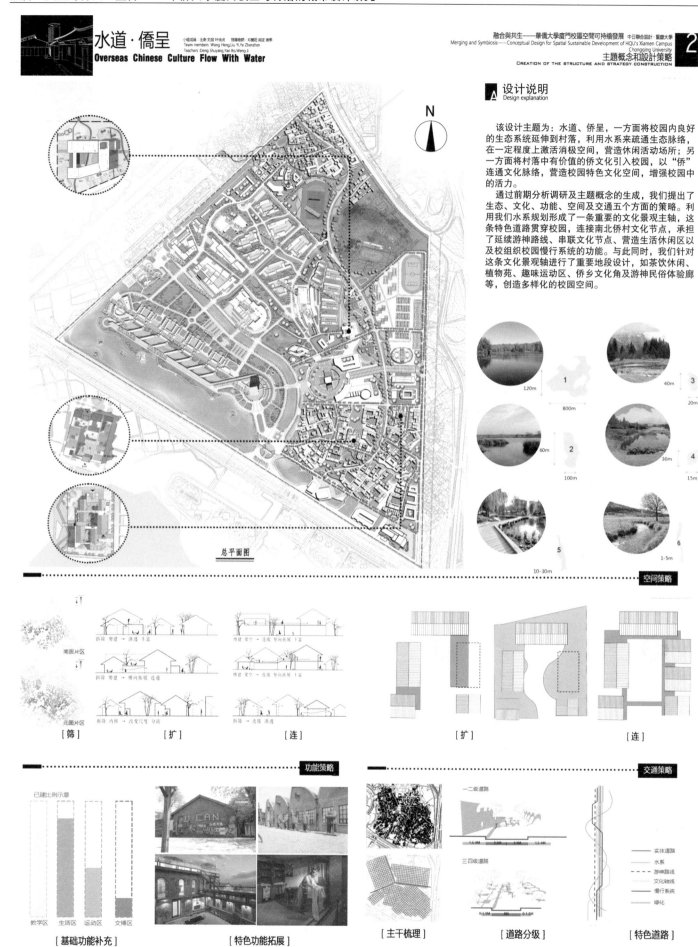

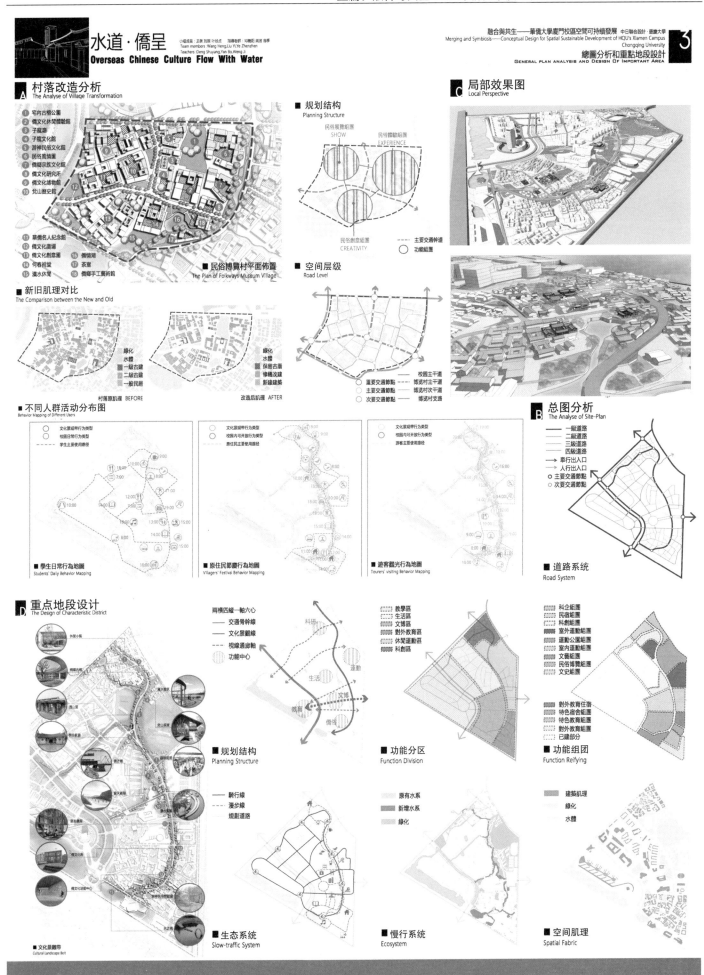

融合·共生与介入·整合 —— 华侨大学厦门校区与村落的城市设计研究

水道·侨呈
Overseas Chinese Culture Flow With Water

小组成员：王恒 刘屹 叶栋真
Team members: Wang Heng, Liu Yi, Ye Zhenzhen
指导老师：邓蜀阳 阎波 翁季
Teachers: Deng Shuyang, Yan Bo, Weng Ji

融合与共生——华侨大学厦门校区空间可持续发展 中日联合设计·重庆大学
Merging and Symbiosis——Conceptual Design for Spatial Sustaiable Development of HQU's Xiamen Campus Chongqing University
游院·侨文化体验休闲中心
Go sightseeing the yard—Overseas Chinese Culture Experience Leisure Centre

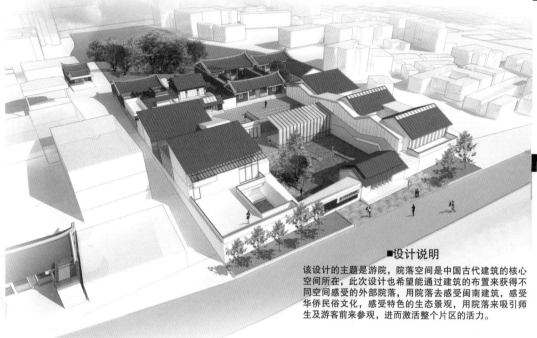

A 设计选址
Site Selection

B 场地简介
Project profile

该设计通过对场内建筑进行筛选，拆除没有价值的建筑，让历史建筑能充分呈现在人们面前，并且也能和周边的建筑及景观资源形成良好的呼应。

设计说明
该设计的主题是游院，院落空间是中国古代建筑的核心空间所在，此次设计也希望能通过建筑的布置来获得不同空间感受的外部院落，用院落去感受闽南建筑，感受华侨民俗文化，感受特色的生态景观，用院落来吸引师生及游客前来参观，进而激活整个片区的活力。

C 形体生成
Form Generation

D 总平面图
General Layout

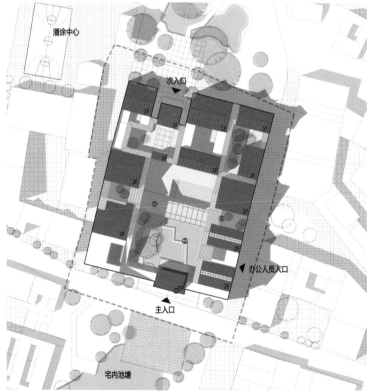

E 院落分析
Courtyard Analysis

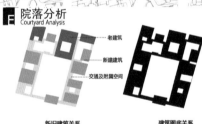

新旧建筑关系 | 建筑图底关系 | 建筑轴线关系
Old and new buildings | Relationship of building figure | The axis in the building

轴线上的三进院落 | 院落间关系 | 空间渗透
Three courtyards on the axis | Relationship between the courtyard | Space penetration

F 院落透视图
Perspective drawing of the courtyard

G 院落设计分析
Courtyard Design Analysis

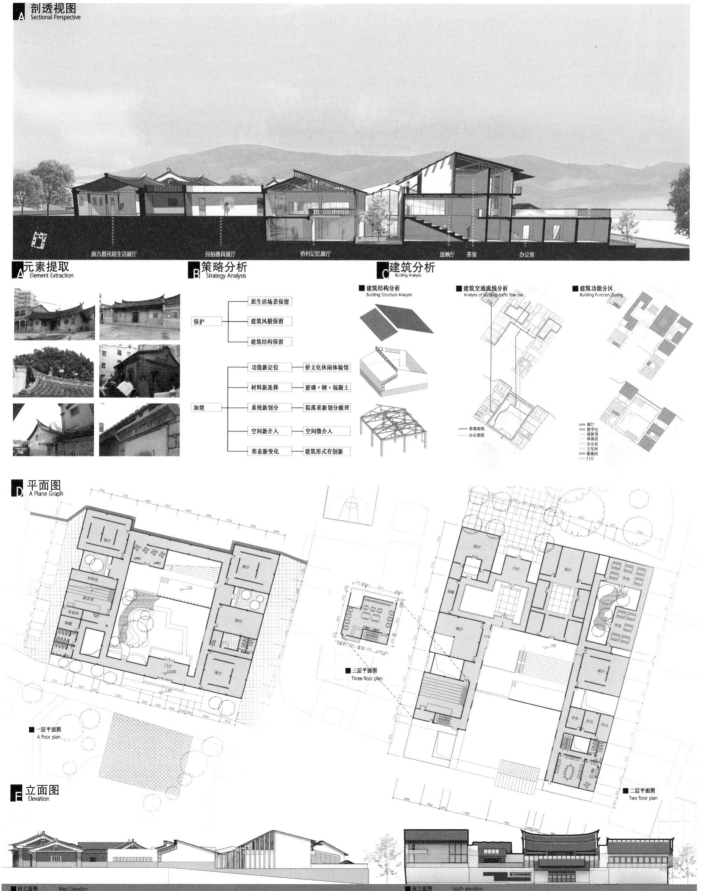

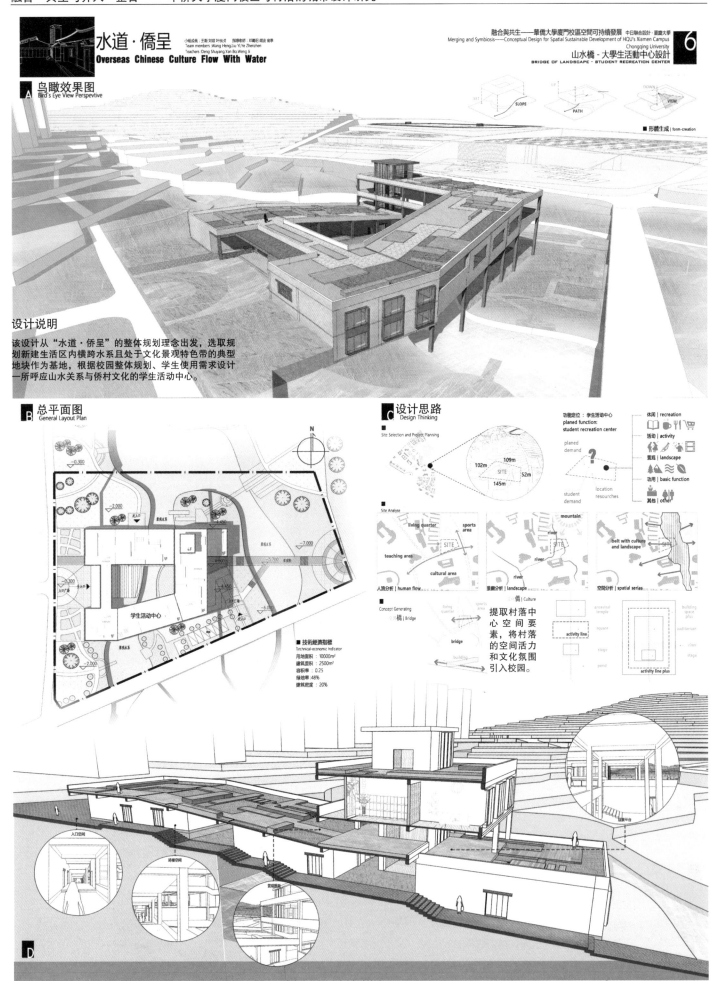

上篇：融合与共生——华侨大学厦门校区校园空间可持续发展概念设计

水道·侨呈
Overseas Chinese Culture Flow With Water

小组成员：王衡 刘弩 叶桢桢
Team members: Wang Heng, Liu Yi, Ye Zhenzhen
指导老师：邓蜀阳 阎波 翁季
Teachers: Deng Shuyang, Yan Bo, Weng Ji

融合與共生——華僑大學廈門校區空間可持續發展 中日聯合設計·重慶大學
Merging and Symbiosis——Conceptual Design for Spatial Sustainable Development of HQU's Xiamen Campus
Chongqing University

山水橋 - 大學生活動中心設計
BRIDGE OF LANDSCAPE · STUDENT RECREATION CENTER

A 各层平面图
Plans of Each Level

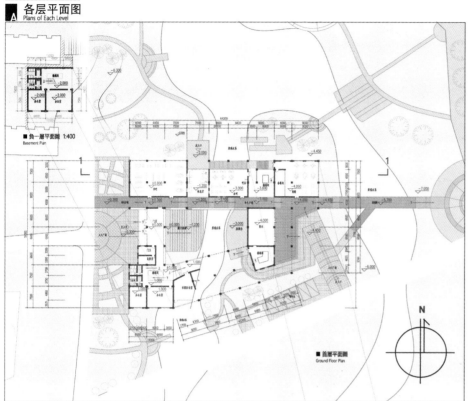

■ 负一层平面图 1:400
Basement Plan

■ 首层平面图
Ground Floor Plan

B 局部效果图
Local Perspective

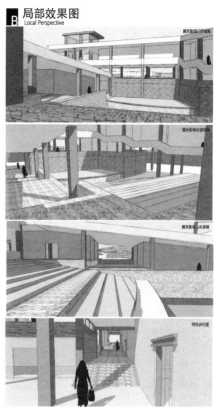

■ 二层平面图
Second Floor Plan

■ 屋顶平面图
Roof Garden Plan

■ 一二层夹层平面图
Middle Floor Plan

C 1-1 剖面图
Profile Map

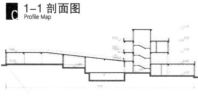

D 立面图
Elevations

■ 东立面图
East Elevation

■ 北立面图
North Elevation

E 分层轴测图
Axonometrical Drawing

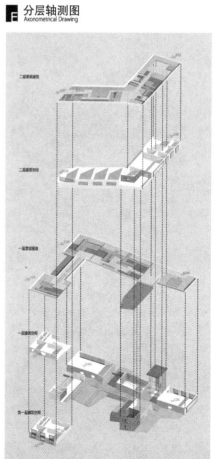

融合·共生与介入·整合 —— 华侨大学厦门校区与村落的城市设计研究

引侨衔村——以侨乡文化为导向的校村融合模式研究
Overseas Chinese hometown culture oriented school village fusion model research

LEAD CHINESE BIT VILLAGE

指导老师：邓蜀阳 阎波 翁季
设计者：陈启泉 曾橙 朱宇诗

教师评语：

作品发现了华侨大学厦门校区与周边村落之间共有的文化基质，提出"引侨衔村"这一概念，希望通过"侨文化"这一主题来建立"校村共生轴"——以校村道路系统作为实轴，以完善的校园功能完善接侨节点，以侨乡文化为虚轴，重要各节点串化形成特色鲜明、生态环境优美的新型大学校园。

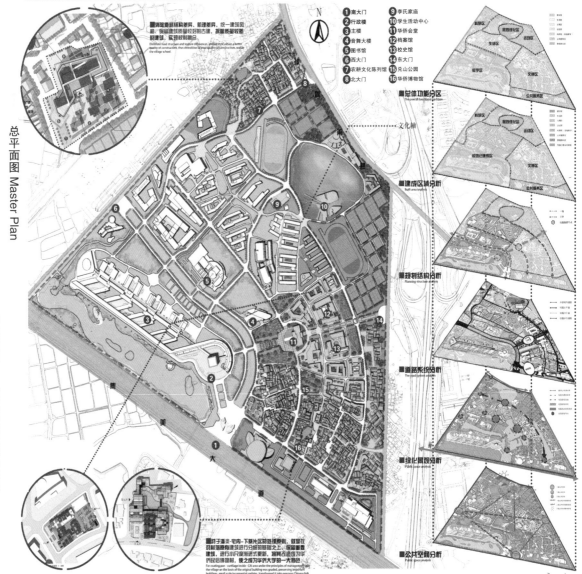

引侨衔村——以侨乡文化为导向的校村融合模式研究

Overseas Chinese hometown culture oriented school village fusion model research

上篇：融合与共生——华侨大学厦门校区校园空间可持续发展概念设计

方案介绍——设计的校村融合模式是什么样的？

重点地段总平面图 Key Section Master Plan

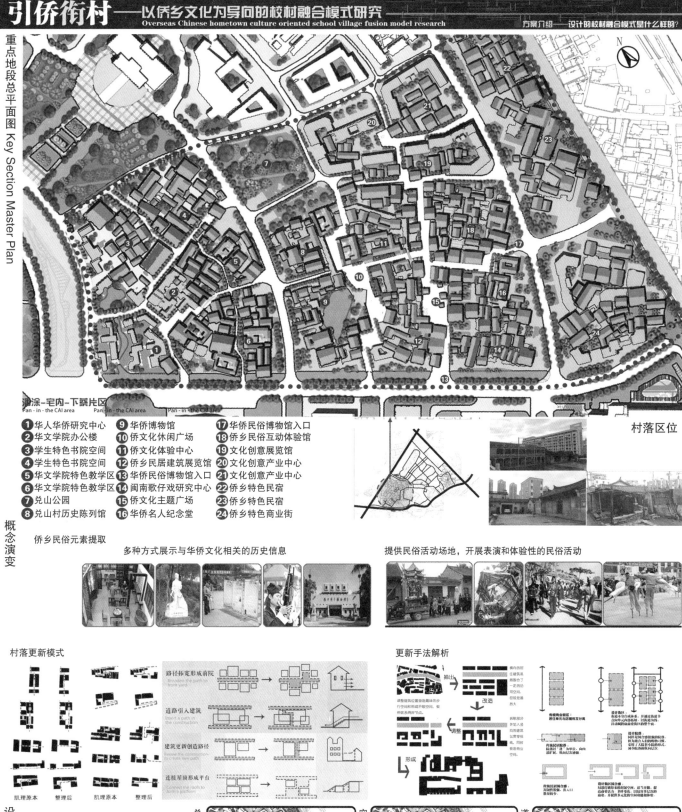

浦涂-宅内-下莲片区
Pan - in - the CAI area

① 华人华侨研究中心
② 华文学院办公楼
③ 学生特色书院空间
④ 学生特色书院空间
⑤ 华文学院特色教学区
⑥ 华文学院特色教学区
⑦ 兑山公园
⑧ 兑山村历史陈列馆
⑨ 华侨博物馆
⑩ 侨文化休闲广场
⑪ 侨文化体验中心
⑫ 侨乡民居建筑展览馆
⑬ 华侨民俗博物馆入口
⑭ 闽南歌仔戏研究中心
⑮ 侨文化主题广场
⑯ 华侨名人纪念堂
⑰ 华侨民俗博物馆入口
⑱ 侨乡民俗互动体验馆
⑲ 文化创意展览馆
⑳ 文化创意产业中心
㉑ 文化创意产业中心
㉒ 侨乡特色民宿
㉓ 侨乡特色民宿
㉔ 侨乡特色商业街

村落区位

概念演变

侨乡民俗元素提取

多种方式展示与华侨文化相关的历史信息

提供民俗活动场地，开展表演和体验性的民俗活动

村落更新模式

肌理原本　整理后　　肌理原本　整理后

- 路径拓宽形成前院 Broaden the path to a front yard
- 道路引入建筑 Insert a path in the construction
- 建筑更新创造路径 Remove the construction to create new path
- 连接屋顶形成平台 Connect the roofs to form a platform

更新手法解析

抽出 → 改造 → 调整 → 形成

设计说明

对于此区域的设计原则，就是既可以与学校部分功能融合，也可以考虑对外开放，并结合适度的旅游商业开发，打造为华侨大学的一大特色。

通过前期调研和分析，将村落内的建筑分为三类：全部保留、适当修缮予以保留、可拆除等。在此基础上植入新的功能体量，在室外空间铺地的处理上延续原有村落的肌理结构，显示对原有村落的态度，延续村民对村落使用空间的回忆。

总体功能分区 / 空间结构分析 / 道路系统分析 / 绿化景观分析 / 公共空间分析 / 重要节点分析

Merging & Symbiosis

引侨衔村——以侨乡文化为导向的校村融合模式研究
Overseas Chinese hometown culture oriented school village fusion model research

鸟瞰图 Bird's Eye View

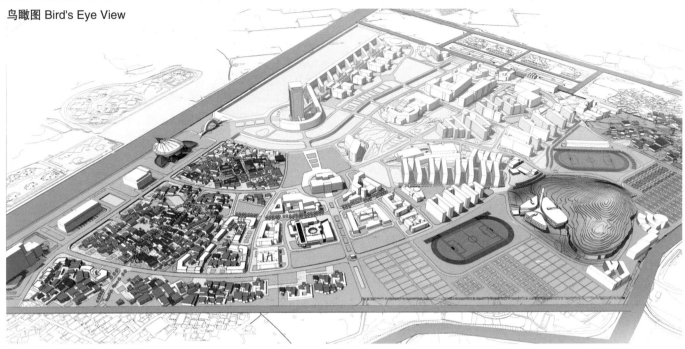

活动时间 The Activity Time

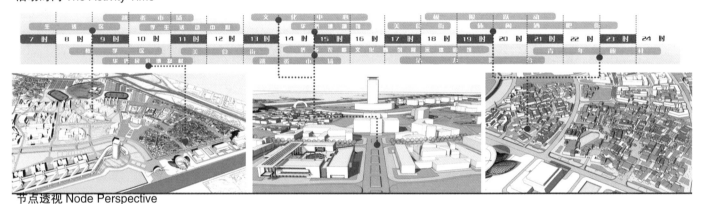

节点透视 Node Perspective

叙事场景构建 To Construct Narrative Scene

cut 1: 引入、到达
镜头拉远，观建筑群体之势。入口艺术灯光装置，引导人群。

cut 3: 循行
在封闭和半封闭的内街和园林中，感受建筑的间隙。我看你，你看我，都是风景。

cut 5: 再循行
将室内和室外颠倒关系处理，制造进行中的新的空间体验。

cut 2: 回望
潜望镜的纵深感，高宽比极致的空间，头顶星云灯束。

cut 4: 升起
山重水复，立面的剖面化的表达，强化建筑体验的场景感，制造观察者和移动者的距离。

cut 6: 观去
在悬挑的"戏台"上回看来的路径，俯瞰场地。

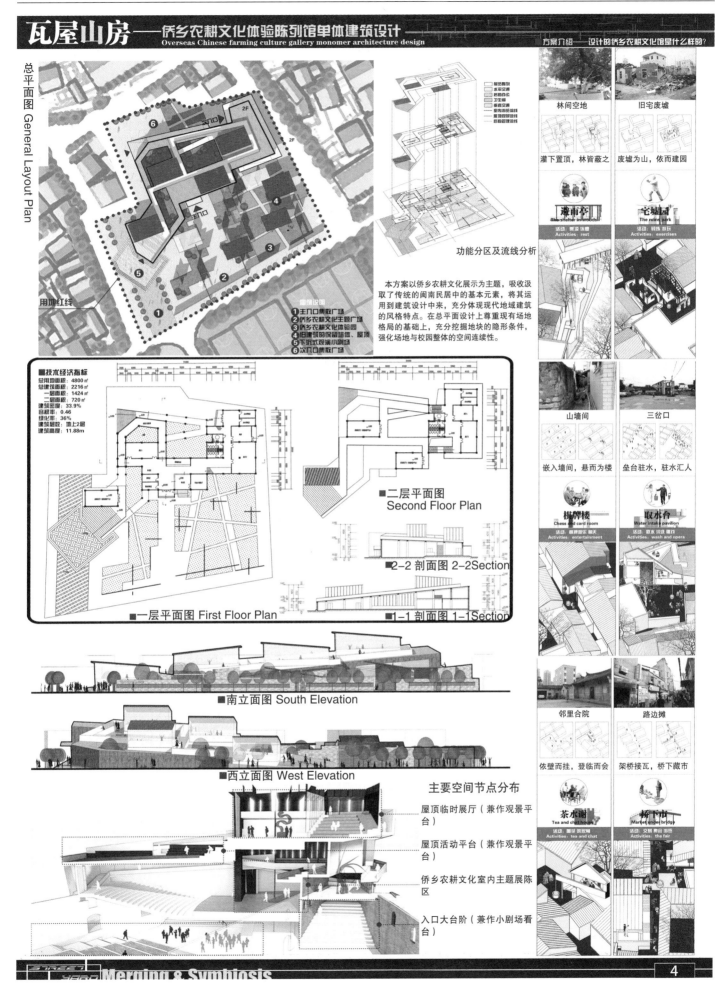

融合·共生与介入·整合——华侨大学厦门校区与村落的城市设计研究

瓦屋山房——侨乡农耕文化体验陈列馆单体建筑设计
Overseas Chinese farming culture gallery monomer architecture design

整体鸟瞰图 The Bird's Eye View

建筑区位 Location

建筑选址位于华侨大学厦门校区西北部,属于校园与城中村的过渡地带。

前期策划 Planning

建筑材料选择

在建筑表皮材料的选择上,尽可能地以吸收当地的本土材料元素,充分体现现代闽南地域建筑的显著特点。

木片墙 Wood / 原有砖墙 Brick
金属栏杆 Metal / 水泥围墙 Cement
新建建筑格栅立面 Grating

概念生成 Concept

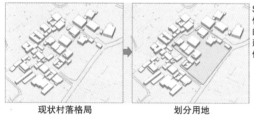

现状村落格局　　划分用地

STEP1:
依据现状村落格局的肌理形态,确定建筑的实际用地区位。

平整场地　　留出交通空间

STEP2:
确定建筑的实际用地区位后,对用地内的原有建筑进行评估分级。

增加公共空间　　加入半开放空间

STEP3:
在建筑前期策划的基础之上适当增加必要的公共空间,同时结合室外交通空间融入半开放空间。

倾斜屋面　　加入屋顶观景平台

STEP4:
对建筑做倾斜屋面的处理,同时加入屋顶观景活动平台,弱化建筑的体量,强化建筑作为观景构筑物的存在形式。

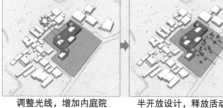

调整光线,增加内庭院　　半开放设计,释放活动空间

STEP5:
在倾斜屋面的基础之上,依据当地的日照分析,调节光纤,在建筑内部增加景观内庭院,满足采光通风要求。

场地景观设计　　形态叠加

STEP6:
对场地内部的景观做进一步细化设计,恢复农耕文化的意向,保留已拆除建筑的墙体以及部分屋顶形式。

上篇：融合与共生——华侨大学厦门校区校园空间可持续发展概念设计

侨乡厝落深——侨文化艺站
Overseas Chinese Cultural Arts station

整体效果展示 Overall Effect Display

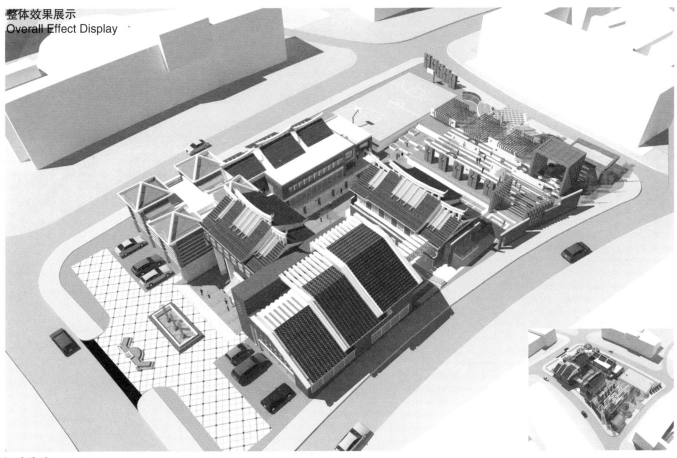

场地选址 Location

建筑单体基地选取在了引侨衔村的关键校园文化带的周边地块，基地内地市平坦，四周都有建成或待建的建筑物。

单体生成 Monomer Generation

基地周边环境分析

西南侧道路为文化带，其他均为校园主干道，因此主入口选择在文化带道路上，次入口在东北侧校园主干道。

文化站的建筑序列和院落应当呈现出纵向布置。基地内有条件设置大量的绿地和广场。

为了应对西南和东北方向建筑的压迫，西南方向要留出入口广场，东北方向设置休憩艺园。

地块划分

鉴于基地具有长进深的特点，将基地纵向划分为5段。

取中间两段作为建筑占地。

融入侨乡厝落的理念，在中间两段植入L形的冷巷，打造出半开放的庭院空间。

寓意诠释 Interpretation of Meaning

通过观察和调研，从闽南民居的排列上得到了灵感，对其纹理重新进行提取和排布，形成了下面的图底形式。

纹理中所包含的黑白灰关系在中国传统水墨抽象画中也非常明显，将水墨画中与前面纹理有关的地方提取出来作为纹理排列依据。

最终将诸多中国南方文化元素交融在一起，形成了富有变化又暗含一定红白灰规律的立面。

取古之大厝形式，融现代建筑结构，营造出新闽南风格的文化站形象展示界面。

透视展示

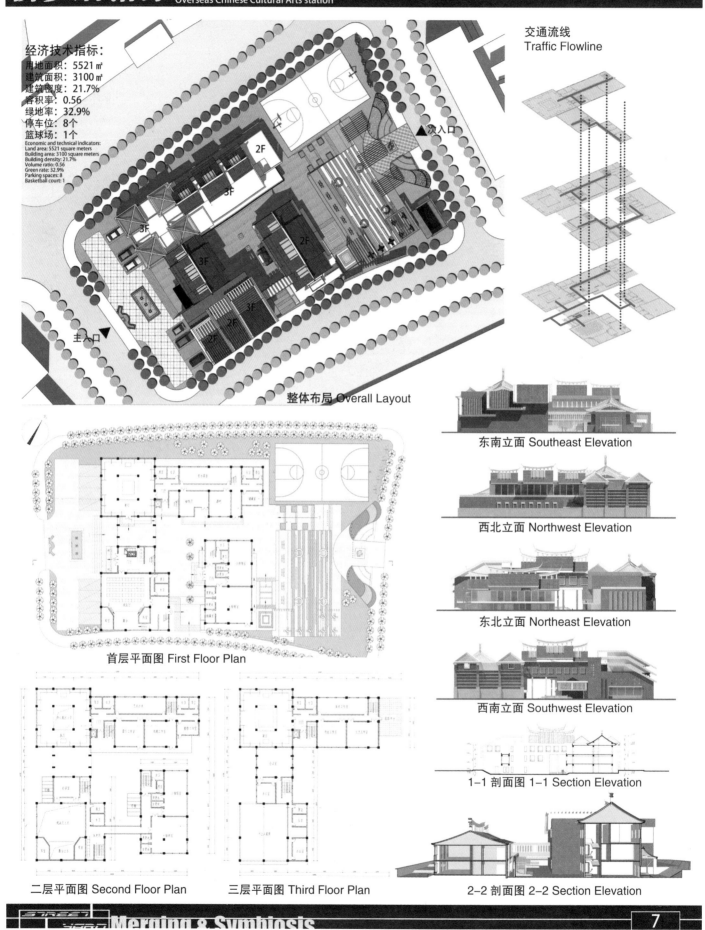

上篇：融合与共生——华侨大学厦门校区校园空间可持续发展概念设计

桥意侨忆——华人华侨博物馆单体建筑设计
ARCHITECTURAL DESIGN OF OVERSEAS CHINESE MUSEUM

鸟瞰图 Aerial View

前期分析 Up-Front Analysis

总平面图 General Layout

首层平面图 Ground Floor Plan

1. 藏品库 STOREROOM
2. 藏品工作室 STUDIO
3. 学术活动室 ACTIVITY ROM
4. 研究室 LABORATORY
5. 办公室 OFFICE
6. 馆长室 CURATOR ROOM
7. 展厅 EXHIBITION ROOM

二层平面图 Second Floor Plan

1. 门厅 HALL
2. 门卫室 MANAGEMENT ROOM
3. 讲解员休息室 NARRATOR LOUNGE
4. 休息室 REST ROOM
5. 纪念品商店 STORE
6. 展厅 EXHIBITION ROOM
7. 制作间 MAKING ROOM
8. 茶室 TEAHOUSE
9. 临时展区 TRANSIT EXHIBITION AREA

概念生成 Concept

闽南大厝建筑形式 Traditional architectural form

提取建筑形式 Extract building form

整理形态组合形体 Composite shapes

呼应场地中水的存在，以桥联系建筑，形成闽式的大厝建筑布局。
Use the bridge to contact buildings.

以桥来隐喻侨信，呼应城市设计主题即侨文化关注。
Use the bridge to metaphor the culture of overseas Chinese.

对传统的大厝布局进行现代转换推局。
Echoing the layout of traditional architecture.

西立面图 West Elevation

北立面图 North Elevation

中国科学院大学
UNIVERSITY OF CHINESE ACADEMY OF SCIENCES

目录 CATALOG

上篇：融合与共生——华侨大学厦门校区校园空间可持续发展概念设计

学生团队 Student Team

陈一山　卢汀莹　彭宁　高林　丁沫

王文慧　李佩蓉　李加丽　边如晨　宋修教

融合·共生与介入·整合 —— 华侨大学厦门校区与村落的城市设计研究

求同存异
SEEKING UNITY WHILE RETAINING UNIQUENESS

指导老师：张路峰　课程助教：彭相国
设计者：王文慧　宋修教　陈一山

概念设计 / SEEKING FOR UNITY WHILE RETAINING UNIQUENESS

指导老师：张路峰　Instructor: Zhang Lufeng　2016-06-20

目标立场——基于活动的"求同存异"
STANDPOINT : SEEKING FOR UNITY IN ACTIVITY

- DIFFERENT ACTIVITIES
- ALL ACTIVITIES
- SIMILAR ACTIVITIES

设计策略——引入触媒
STRATEGY : ACCELERANT

现状　　　　手法1　　　　手法2　　　　手法3
校、村矛盾激烈　校吞并村　　放任自流　　引入第三方——触媒协调

function : museum about Huaqiao
form : scatters

触媒两要素
置入功能：侨文化博物馆
置入形式：散点状

触媒功能
FUNCTIONS OF ACCELERANT

侨文化展览　节日活动　村落研究　研究　祭祖　游玩　娱乐　文艺活动　休憩

触媒形式
FORM OF ACCELERANT

触媒（博物馆）面积指标　　其他活动所需面积

打散
break

第一步：放入场地　STEP 1 : put all into site
第二步：打散体块　STEP 2 : break block
第三步：小体块重组　STEP 3 : scatters recombine
第四步：按秩序排列　STEP 4 : arrange into order

点阵排布
ORDER OF SCATTERS

对比常见的六边形、三角形、正方形网格，后者辨识性和引导作用更强

网格方向1：天马山到杏林湾连线
网格方向2：集美大道沿线
网格方向确定

尺度1：间距100m 点太密、太多；点与点之间环境变化很小
尺度2：间距200m 点数量适中；点与点之间环境变化较大；可达性较好
尺度3：间距300m 点数量过少；可达性较差

点阵调整
ADJUSTING THE ORDER

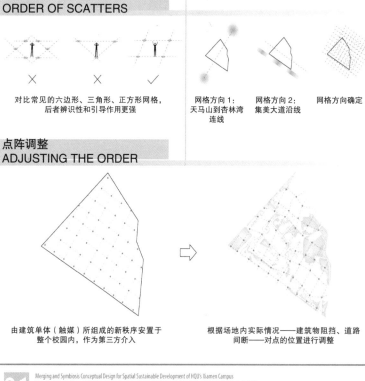

由建筑单体（触媒）所组成的新秩序安置于整个校园内，作为第三方介入

根据场地内实际情况——建筑物阻挡、道路间断——对点的位置进行调整

对场地内所有触媒点进行功能主题分类，使整个系统更加整体

龙舟文化
村社中心
民俗展示
名人事迹

教师评语：

　　该设计组调研发现虽然场地内人群在活动时间、活动范围、语言、年龄、教育背景等方面均有较大差异，但是人群基于华侨文化的活动是共通的。因此，华侨文化的相关活动为融合提供一种可能性。作品基于触媒理论，以形似拉维莱特公园的设计手法将均布点状的各类华侨文化博物馆植入场地。这些博物馆不仅满足了华侨文化与历史的保存与展示，其本身也成为场地内的重要节点将校园空间秩序加以织补与重组。

01 融合与共生·华侨大学厦门校区校园空间可持续发展概念设计
Merging and Symbiosis Conceptual Design for Spatial Sustainable Development of HQU's Xiamen Campus
2016年五校联合毕业设计终期汇报

UCAS

上篇：融合与共生——华侨大学厦门校区校园空间可持续发展概念设计

城市设计 / SEEKING FOR UNITY WHILE RETAINING UNIQUENESS 求同存异

王文慧 Wang Wenhui　宋修教 Song Xiujiao　陈一山 Chen Yishan
指导老师：张路峰　Instructor: Zhang Lufeng　2016-06-20

总平面图 / Master plan

- C4: History Archives
- B6: Xiheng Museum
- C6: Culture Amusement Park
- C3: Water Front Museum
- E6: Family Li's temple Museum
- D3: Exhibition Bridge
- E6: Memorial Museum
- E3: Handicraft Workshop
- F4: Garden Tower
- E1: Dragon Boat Museum
- F3: Creative Museum
- H3: Xianming Temple Museum

城市设计导则 / Urban Design Guideline

1. The buildings divided into four types cultural heritage activated protection, folklore studies, history shows
2. Each building want to melt surrounding crowd, in order to play a catalytic mechanism
3. The function of building comes from surrounding space within range of 200m×200m
4. In materials, every single building has at least a facade of "a brick into the stone."
5. In structure, each single building need to "3m, 6m, 9m" modulo framework adopted
6. In space, every single building shall have a framed space, enclosed in any manner whatsoever
7. To be considered between the guidance system and the monomer

Merging and Symbiosis Conceptual Design for Spatial Sustainable Development of HQU's Xiamen Campus
融合与共生:华侨大学厦门校区校园空间可持续发展概念设计
2016 年五校联合毕业设计终期汇报

融合·共生与介入·整合 —— 华侨大学厦门校区与村落的城市设计研究

单体设计 — SEEKING FOR UNITY WHILE RETAINING UNIQUENESS 求同存异

陈一山 Chen Yishan
指导老师：张路峰
Instructor: Zhang Lufeng   2016-06-20

总平面图 /Master Plan

显明宫博物馆
Xianming-temple Museum (H3) located in the eastern part of the campus into the village street, close to the Temple. The street has become a good business climate, as well as the surrounding area and kindergarten. Surrounding residents are migrant workers. Xianming-temple Museum obvious link -G3,H4,J2 -three other buildings. The museum is good accessibility, setting identification alone the street to guide flow. The Museum could improve the environment of the street and enclose a small plaza for activity.
场地 / Site

总平面图 /Master Plan

创意工坊博物馆
Creative Workshop Museum (F3) located in the eastern part of the campus near the main commercial street. Sides of the street are mostly restaurants and entertainment facilities. Most of the surrounding residents are migrant business people. Creative Workshop Museum contact-E3,E4,G3,G4-four related buildings. The museum is good accessibility, located at the intersection of the street with identification. Creative Workshop Museum will provide a place to melt students and residents.
场地 / Site

市场 /Market

场地分析 / Analysis

去往 D3/ 娱乐/ 餐饮/ 场地分析 /

场地分析 / Analysis

二层 /Second Floor

The Museum could hold community events, history Exhibition and overseas cultural propaganda. It is located in a commercial street so it add a tearoom. Villagers can set the the market. God will also travel there when Festival are coming. Students and villagers could take part in some activities to recognize each other.

首层 /Ground Floor

Creative Workshop Museum provide a place to teaching manual skills, selling handicraft production and doing small research.In the ground floor, it offers space for local craftsmanship inheritors to make and sell production. In the second floor there are many small mobile lab to provide students with self-study and research handicraft. Students and inheritors of the traditional process can be recreated in there.

二层 /Second Floor

生成图解 /Graphic Generation

功能分析 /Functional Analysis

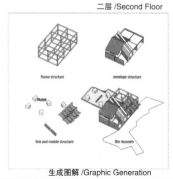
生成图解 /Graphic Generation

03 Merging and Symbiosis Conceptual Design for Spatial Sustainable Development of HQU's Xiamen Campus
融合与共生:华侨大学厦门校区校园空间可持续发展概念设计
2016 年五校联合毕业设计终期汇报

单体设计 / SEEKING FOR UNITY WHILE RETAINING UNIQUENESS 求同存异

陈一山 Chen Yishan 指导老师：张路峰 Instructor: Zhang Lufeng 2016-06-20

总平面图 / Master Plan

葬仪博物馆
Funeral Museum (E6) located in the northern mountains, near the Primary School, revolutionary martyr monument, and Maoshan-temple. The Hill is local villagers ancestral burial place. Now the hill is also a forest garden.
Funeral Museum contact-D5, D6, E5, F6-four related buildings. The museum is a good ecological environment, between the burial area and the park area. Due to the remote location, the museum is open-air museum, without guarded.
场地 / Site

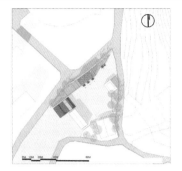

家庙博物馆
Family Temple Museum (D5) is located near the Duishan commune and the Family Li's temple. The family temple is now the demarcation of villages and schools.
Family temple museum contact-C5, C6, D6, E4, E5-five related buildings. The museum's accessibility is very good, located in the only way the villagers and the student out of school. The museum will strengthen the center of the village a sense of enclosure.
场地 / Site

场地分析 / Analysis

场地分析 / Analysis

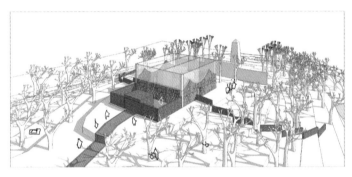

首层 / Ground Floor

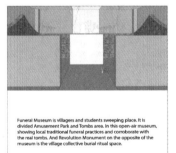
Funeral Museum is villagers and students sweeping place. It is divided Amusement Park and Tombs area. In this open-air museum, showing local traditional funeral practices and corroborate with the real tombs. And Revolution Monument on the opposite of the museum is the village collective burial ritual space.

首层 / Ground Floor

Li's temple museum is the functions of the supplement of the Duishan Commune and Lee temple. It provide rehearsal and changing places. Museum show the history of temple. The village center as a place for the villagers daily activities. The new building will enhance the quality of the surrounding to attract students to come here to visit and study.
二层 / Second Floor

功能分析 / Functional Analysis

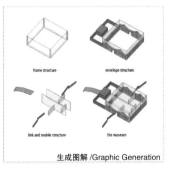
生成图解 / Graphic Generation

功能分析 / Functional Analysis

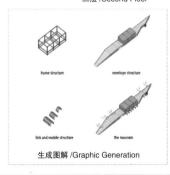
生成图解 / Graphic Generation

融合·共生与介入·整合——华侨大学厦门校区与村落的城市设计研究

单体设计 / SEEKING FOR UNITY WHILE RETAINING UNIQUENESS 求同存异

王文慧 Wang Wenhui

指导老师：张路峰 Instructor: Zhang Lufeng 2016-06-20

总平面图 /Master Plan

点阵位置 /Lattice Position

C4 History Archives:
History Archives (C4) located in a neglected lawn, surrounding by dormitories, canteen, College buildings. The establishment of the underground history archives, not only would retain the green area, but increase the use of the green, as a landscape sketch.

总平面图 /Master Plan

点阵位置 /Lattice Position

C4 History Archives:
History Archives (C4) located in a neglected lawn, surrounding by dormitories, canteen, College buildings. The establishment of the underground history archives, not only would retain the green area, but increase the use of the green, as a landscape sketch.

场地现状 /Site Status

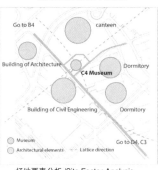
场地要素分析 /Site Factor Analysis

场地现状 /Site Status

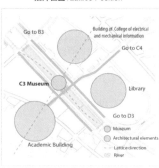
场地要素分析 /Site Factor Analysis

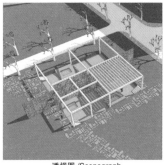
透视图 /Scenograph

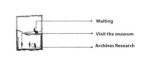
活动分析 /Activity Analysis

透视图 /Scenograph

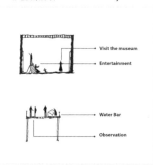
活动分析 /Activity Analysis

透视图 /Scenograph

透视图 /Scenograph

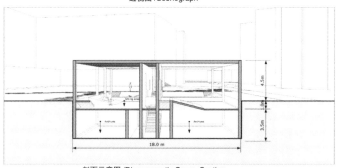
剖面示意图 /Diagrammatic Cross-Section

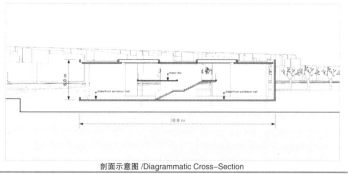
剖面示意图 /Diagrammatic Cross-Section

Merging and Symbiosis Conceptual Design for Spatial Sustainable Development of HQU's Xiamen Campus
融合与共生:华侨大学厦门校区校园空间可持续发展概念设计
2016年五校联合毕业设计终期汇报

单体设计 / SEEKING FOR UNITY WHILE RETAINING UNIQUENESS 求同存异

王文慧 Wang Wenhui

指导老师：张路峰
Instructor: Zhang Lufeng　2016-06-20

总平面图 /Master Plan

E3 Crafts workshop:

Handicraft workshop (E3) is located at the southeast edge of the University, as a transition region between school and village, providing students with entrepreneurial practice places and some local arts and crafts shops. It is a living museum of Arts and crafts.

点阵位置 /Lattice Position

总平面图 /Master Plan

C6 Cultural amusement park:

Cultural amusement park (C6) is located in the north boundary of the University, adjacent to the sports facilities of campus, kindergarten and a large number of residential buildings. By using the enclosure of green space and dilapidated houses, combined with the functionality of the museum, we intend to provide a place for the children's entertainment and fitness, which would lead more families to understand overseas Chinese culture.

点阵位置 /Lattice Position

场地现状 /Site Status

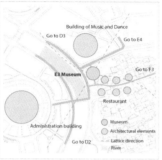

场地要素分析 /Site Factor Analysis

场地现状 /Site Status

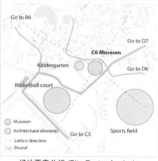

场地要素分析 /Site Factor Analysis

透视图 /Scenograph

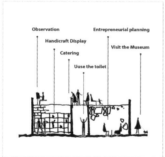

活动分析 /Activity Analysis

透视图 /Scenograph

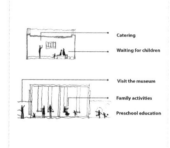

活动分析 /Activity Analysis

透视图 /Scenograph

透视图 /Scenograph

剖面示意图 /Diagrammatic Cross-Section

剖面示意图 /Diagrammatic Cross-Section

融合·共生与介入·整合 —— 华侨大学厦门校区与村落的城市设计研究

单体设计 / SEEKING FOR UNITY WHILE RETAINING UNIQUENESS 求同存异

宋修教 Song Xiujiao
指导老师：张路峰　Instructor: Zhang Lufeng　2016-06-20

总平面 /Master Plan

场地位置 /Location

本设计选址于西衡社村落中心位置附近，功能定位为可以激活村落中心的活态"侨文化博物馆"，采用软性的介入，加了一个楼梯以联系各服务空间，对广场进行铺地设计以增强整体性。

Located at the center of XIHENG SHE, it's main function is a "QIAO Culture Museum", which can activate the center. By some soft methods, we take some measures about rebuilding the pave and adding a staircase.

设计说明 /Introductions

透视图 /Rendering

礼堂正对着戏台，二者之间形成一条公共空间的纵向轴线。
The temple face to the stage, so there exist an axis in the space.

改造位于纵向轴线两侧的两栋废弃小方楼作为服务空间，并将二者联系起来。
Renovate the 2 block-buldings as service space for the axis, and combine them.

在方形篮球场边有一个粮仓和两个古厝，文化价值高但已废弃。
There exist 1 granary and 2 ancient Cuo, whichis of great culture value but abandoned.

根据原有空间关系对铺地进行设计，联系起篮球场和三栋房子，留下活动可能。
Design the pavage to combine them according to the existing space relationship.

场地分析及设计 /Site Analysis And Design

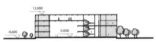

首层 /Ground Floor

1-1 剖面 /1-1 Section

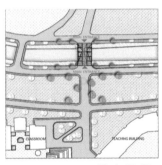

鸟瞰图 /Bird's Eye Perspective

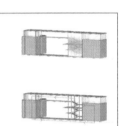

将原有空间进行梳理，明确服务轴和主空间轴的关系为正交。
Analysing the space relations, and we make an cross between service axis and space axis.

由两个方向的建筑肌理延伸线，决定新设计建筑的外轮廓。
We design the outline of the new building according to the exsting culture texture.

新设计的建筑为一个交通核，其主要意义在于将两侧服务空间联为一个整体。
The small house is used as a transportation core, which can combine the service spaces.

交通核结合铺地和飘廊，将两栋房子联系为一个立体服务环。
The transportation core, the pavage, the corridor, and 2 houses form a vertical circle.

建筑分析及设计 /Architecture Analysis And Design

总平面 /Master Plan

场地位置 /Location

本设计选址于连通校园生活区和教学区的一个桥上，功能定位为景观桥，兼具体憩博览功能。在统一的框架基础上，设计一个具有一定动感的景观桥，作为激活周边区域的一个节点。

Located at a bridge connecting the living quaters and teaching quaters, it's defined as a landscape bridge, which have some resting area. Based on a uniformed framework, we design a bridge for activate the surrounding area.

设计说明 /Introductions

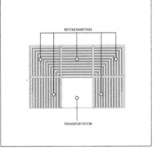

建筑体量 /Building Volume　　建筑分析 /Building Analysis

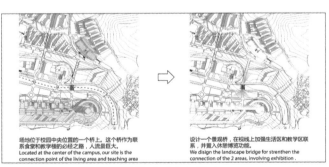

场地位于校园中央位置的一个桥上。这个桥作为联系食堂和教学楼的必经之路，人流量巨大。
Located at the center of the campus, our site is the connection point of the living area and teaching area.

设计一个景观桥，在视线上加强生活区和教学区联系，并置入休憩博览功能。
We design the landscape bridge for strenthen the connection of the 2 areas, involving exhibition.

场地分析及设计 /Site Analysis And Design

融合与共生·华侨大学厦门校区校园空间可持续发展概念设计
2016年五校联合毕业设计终期汇报

UCAS

单体设计 / SEEKING FOR UNITY WHILE RETAINING UNIQUENESS 求同存异

宋修教 Song Xiujiao
指导老师：张路峰
Instructor: Zhang Lufeng 2016-06-20

上篇：融合与共生 —— 华侨大学厦门校区校园空间可持续发展概念设计

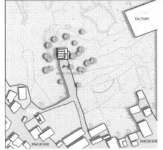
总平面 /Master Plan

场地位置 /Location

本设计选址于下蔡社北侧的空旷田野处，功能定位为较为完整独立的"侨文化博物馆"。主要功能为环绕中庭布置的展览空间，材料上采用了"出砖入石"作为对村落的回应。

This building will be located at open country farmland, thus it can be designed into a complete museum. The center courtyard is surrounded by exhibition space.

设计说明 /Introductions

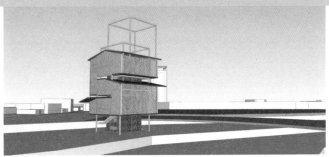

处大片农田之间，周边空旷，景色优美；可相对自由置入完整博览建筑。
Surrounding with rural farmland, the site has got a wonderful scenery.

和村落遥相呼应，建筑设计应考虑村落功能及形态等关系。
Xiacai She located not far from our site, so we should consider the morphology of the country.

场地分析及设计 /Site Analysis And Design

本设计的基本空间原型来自闽南地区传统民居的空间构成，即中间为庭院，前后两侧为厝，左右两侧为护厝的围合式布局。
The space prototype derives from the local folk house. Central yard is surrounded by the rooms.

建筑分析及设计 /Architecture Analysis And Design

剖面分析 /Section Analysis

建筑体量 /Arch Perspective

建筑体量对比推敲 /Dissimilation

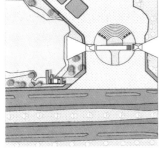
总平面 /Master Plan

场地位置 /Location

本设计选址于校园正门侧面和湖面相邻处，功能定位为具有博览功能的景观小品。设计上采用连续丝带的意向，结合外框架，营造休憩等候空间。

This building locate right beside the campus' main gate, it's actually a landscape architecture for resting and exhibition. Combined with the framework, we make it look like a silk ribbon.

设计说明 /Introductions

透视图 /Rendering

亲水平台/water-side terrace
入口引导/main entrance
完整墙面隔绝噪音/a complete wall keeps it from the sound noise

紧邻大片水面，景观资源优良
Located aside the water, it has wonderful scenery.

于校园正门外侧，人流量巨大
Located outside the main gate of the campus, out site has lots of people pass by.

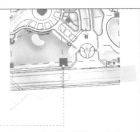
紧邻集美大道，受交通干线噪音影响较大，也位于学校与外界社会的边界
Aside the Jimei Avenue, the site is strongly influenced by the huge transportation.

场地分析及设计 /Site Analysis and Design

融合·共生与介入·整合 —— 华侨大学厦门校区与村落的城市设计研究

顺水推船 TROJAN

指导老师：张路峰 课程助教：彭相国
设计者：高林 李加丽

TROJAN 顺水推船

基地现状 Current Situation

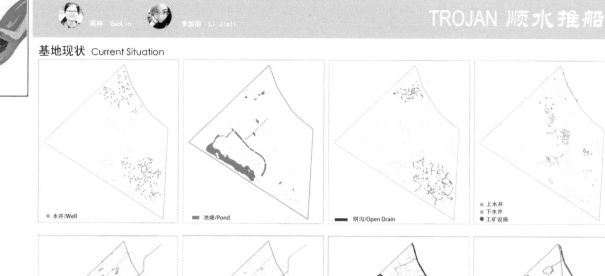

设计策略 Design Strategy

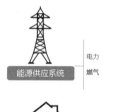

借完善村落基础设施为手段，赋予附属建筑以校园所需功能，达到校村融合的做法——结合基础设施设置"设施附属建筑"（如配电室、污水处理厂等）—结合相关学校内部功能（文博馆、实习基地）—建设师生便利活动中心。围绕村落更新主题对"基础设施"相关方面进行了资料调研，确定设计切入的相关方面。选择原则1、亟待改进 2、可承载学校功能，并且通过谁的改造可以带动其他相关设施的带入。

The infrastructure is quite different between school and the village. So the school can build some the affiliate architecture of the infrastructure, then add some school functions into the buildings, in this way, school and village can use them together.

教师评语：

校园总体规划原有理念是秉承生态优先，先规划了水系，而后形成校园。该作品延续了总规划理念，以水系相关基础设施改造和利用作为校区处理与排水设施及有机整合植入基础设施所需的功能，不但完善了村落基础设施条件，又满足了学校功能扩展的需求，达到双赢的目的。

策略深入 Advanced Strategy

带动其他基础设施/ other infrastructure

增设学校功能/ addition to school function

水系方案 Design Scheme

污水厂、人工湿地置入　　污水排水管道设计　　恢复部分水道连通现状　　雨水排水系统

上篇：融合与共生 —— 华侨大学厦门校区校园空间可持续发展概念设计

TROJAN 顺水推船

高林 GaoLin　李加丽 Li Jiali

方案深入　Advanced Scheme

道路系统　环卫系统　电力系统　收发系统

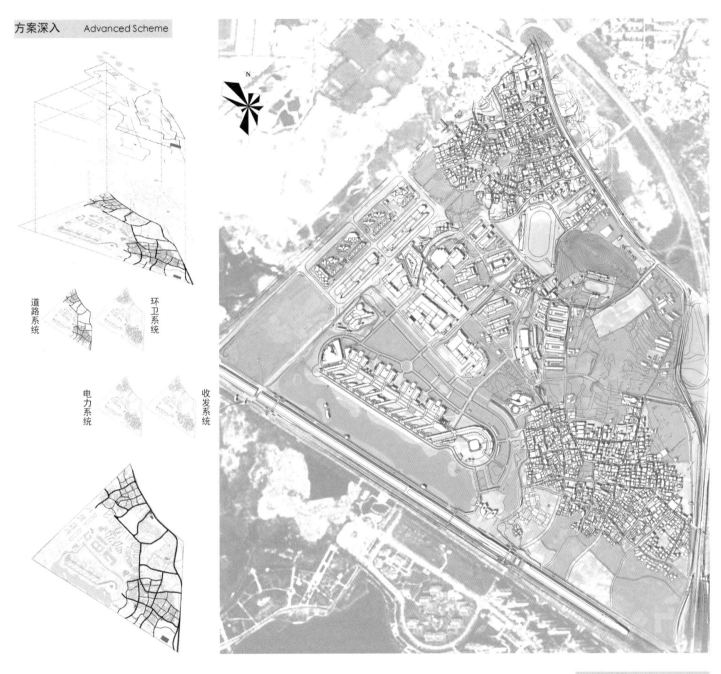

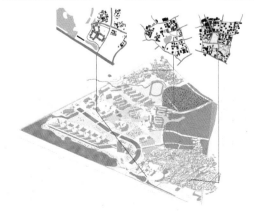

Museum and Sewage treatment　Student Creative Exhibition and Waste Collection　Community center and express mail Electric Room

方案总图　General Plan

总平面图设计将整个校村基地作为整体考虑，配套与面积相当的市政系统，并结合学校对改造点进行选址与定性，最终实现服务校园、城市、村庄为一体。

General plan design regarded the school and village as a whole, with the appropriate municipal supporting system, combined with the situation of the school, finally realize the anastomosing of the village, the campus and the city

融合·共生与介入·整合 —— 华侨大学厦门校区与村落的城市设计研究

高林 GaoLin　李加丽 Li Jiali

TROJAN 顺水推船

博物馆&污水处理厂设计
Museum & Sewage Treatment

由于污水处理的要求，选择基地内地势较低且靠近学校内水再生处理的位置作为村内初次污水处理的基地，由此将水联系校村，实现融合。

Because of the requirement of the sewage treatment, we choose the place which is low lying and closed to the water regeneration treatment as the location of the base.

总平面图 Site Plan

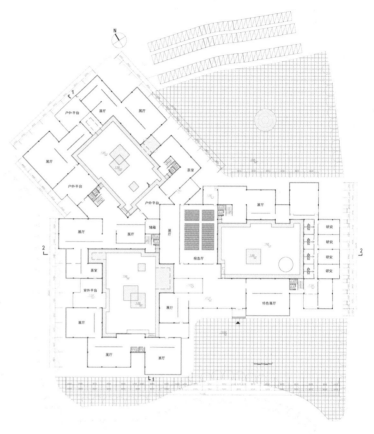

首层平面图 First Floor Plan

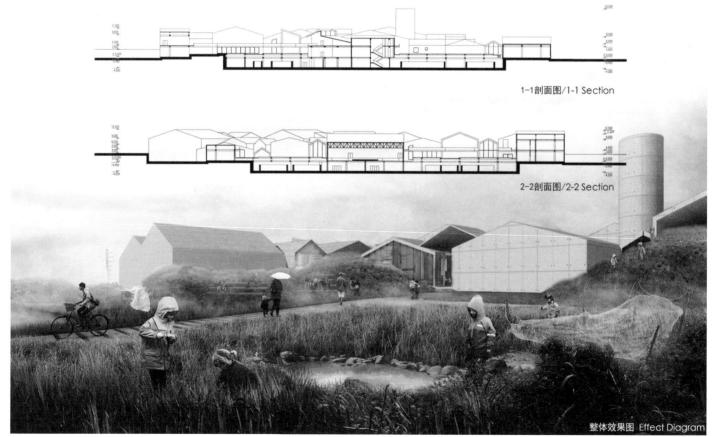

1-1剖面图/1-1 Section

2-2剖面图/2-2 Section

整体效果图 Effect Diagram

上篇：融合与共生——华侨大学厦门校区校园空间可持续发展概念设计

TROJAN 顺水推船

高林 GaoLin　　李加丽 Li Jiali

形体生成 Generation Analysis

地下一层平面图 Basement Floor

交通流线 Traffic Streamline

二层平面图 Second Floor

节点效果 Court Perspective

节点效果 Court Perspective

立面图 Elevation

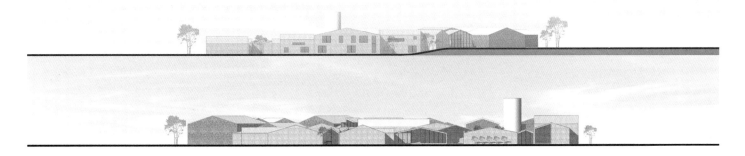

上篇：融合与共生——华侨大学厦门校区校园空间可持续发展概念设计

TROJAN 顺水推船

快递收发室 & 活动中心设计
COMMUNITY CENTRE DESIGN

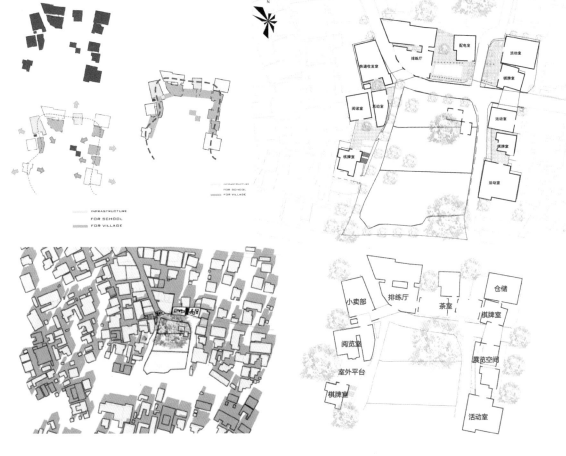

活动中心单体依附于"邮政收发系统"中的"快递收发室"及"能源供应系统"中的"配电室"设置，借由原有建筑的肌理置入预设功能，并加建一些室外连廊将"小空间"串联形成新的活动场域，立面选型借由"古榕公园"选择树形立面，用统一的表皮形式在二层高度处连接建筑，营造景观良好、空间丰富的校村活动中心。

The community centre design rely on the postal transceiver system itself express mailroom and energy supply system in the default function of original building, and build some outdoor nest will "Small Space" series into a new field of the domain, Elevation selection by "Banyan Park" to tree facade, in the form of a unified epidermis connected buildings on the second floor height. Build good landscape and rich apace school village activity center.

融合·共生与介入·整合 —— 华侨大学厦门校区与村落的城市设计研究

出砖入石 EMBEDDING

指导老师：张路峰 课程助教：彭相国
设计者：边如晨 卢汀滢

总体设计

 边如晨 Bian Ruchen
 卢汀滢 Lu Tingying

EMBEDDING 出砖入石

指导老师：张路峰 2016-6-20

项目区位 /Site Location

 中国-厦门 China-Xiamen
 厦门-集美区 Xiamen-Jimei
 集美区-华侨大学厦门校区 Jimei-Hua Qiao University

分区定位 Sections　　新的校门 New Gates　　跳通道路 streets

问题与现状 /Problems&Current Situations

在学校 /In School　　在村庄 /In Village

 闲置的绿地 /Empty Greenland　　缺少文娱设施 /Lack Amusement　　 缺少管理 /Lack Of Management　　随处可见的废墟 /Ruins Everywhere

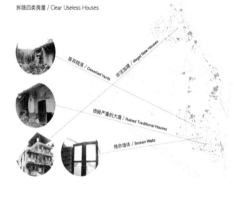

拆除四类房屋 / Clear Useless Houses
废弃院落 / Deserted Yards
非法加建 / Illegal New Houses
损毁严重的大厝 / Ruined Traditional Houses
残余墙体 / Broken Walls

在村校中分别展入对方所需功能
Insert New Functions Into Both Sides

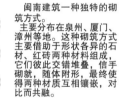

闽南建筑一种独特的砌筑方式。
主要分布在泉州、厦门、漳州等地。这种砌筑方式主要借助于形状各异的石材、红砖两种材料组成，它们彼此交错堆叠，信手砌就，随体附形，最终使得两种材质互相镶嵌，对比而共融。

策略 /Strategy – "出砖入石"/Embedding

石头房 Stone House　　砖房 Brick House　　出砖入石的房子 Embedding House

Present Condition → Ideal Condition

Position & Volumes

教师评语：

"出砖入石"是闽南建筑独特的一种建构手段。该作品基于建筑体量视角，将传统村落的小体量建筑与新建校园的大体量建筑分别隐喻为砖和石，并将新与旧、大与小两类建筑在空间上互相渗透，在功能上互相补充，进而引致不同人群的融合。作品对"出砖入石"这一传统的借鉴不仅停留于空间形态上，更是在思想和方法层面获得了一种让两种不同属性的事物协调矛盾、各展其能的启示。

■ Education　■ Sports　■ Office　■ Activity　■ Exhibition　■ Restaurant & Cafe

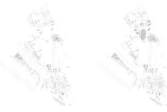

 Merging and Symbiosis Conceptual Design for Spatial Sustainable Development of HQU's Xiamen Campus
华侨大学厦门校区校园空间可持续发展概念设计
2016年五校联合毕业设计终期汇报

 UCAS

融合·共生与介入·整合——华侨大学厦门校区与村落的城市设计研究

单体设计
边如晨 Bian Ruchen

EMBEDDING 出砖入石
指导老师：张路峰 2016-6-20

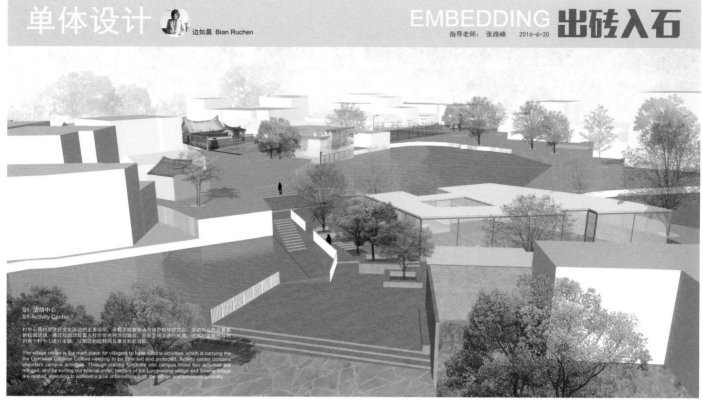

S1- 活动中心
S1-Activity Center

村中心是村民进行文化活动的主要场所，承载着需要继承和保护的华侨文化。活动中心包含重要的校园活动，通过校园功能置入村庄使两种活动融合，并梳理空间秩序，达成对龙围井与西亨两个村中心进行串联，以期达到村校两方惠互利的目标。

The village center is the main place for villagers to have cultural activities, which is carrying the the Overseas Chinese Culture needing to be inherited and protected. Activity center contains important campus activities. Through placing functions into campus, those two activities are merged, and by sorting out spacial order, centers of the Longweijing village and Xiheng village are related, intending to achieve a goal of benefiting both the village and school reciprocally.

场地位置及现状 /Location&Background

场地分析 /Site Analysis

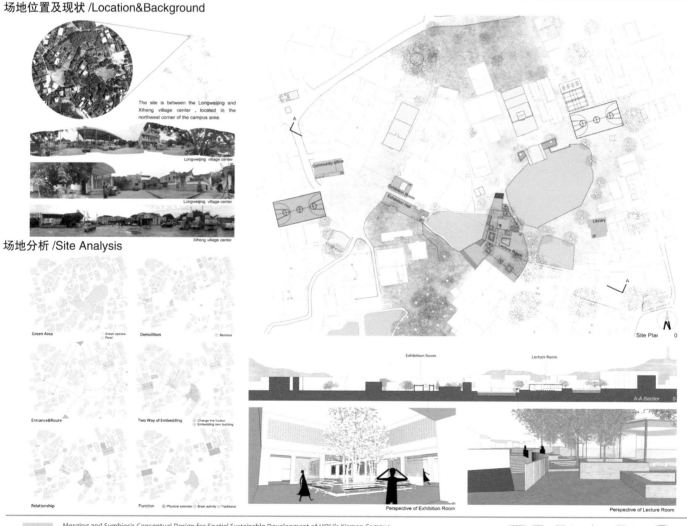

Merging and Symbiosis Conceptual Design for Spatial Sustainable Development of HQU's Xiamen Campus
华侨大学厦门校区校园空间可持续发展概念设计
2016年五校联合毕业设计终期汇报

单体设计 / EMBEDDING 出砖入石

边如晨 Bian Ruchen 指导老师：张路峰 2016-6-20

B1- 综合休息站
B1-Comprehensive Rest Station

该方案为规划中的电瓶车站之一，位于图书馆北侧的大片绿地上，综合休息站一方面为学生们提供等候电瓶车时休息的场所，另一方面也作为可供村民来学校活动的场所，并嵌入村落的服务设施以满足学生对服务设施的需求，同时也为村民们获取利益提供一个机会。建筑形态取自于村落中重要的文化元素，并将村落中拆除的破败房屋的材料进行回收利用，重新形成新的建筑。

The program, located on the northen lawn of the library, is the planning battery station. Comprehensive rest stop, on the one hand, provides students a place of resting when they are waiting for battery cars, on the other hand, acting as a place of acitivites for local villagers, which is inseting into those service facilities of the village to meet both the demands of students and the benefits requirements of villagers. Architectural form is shaped from the important cultural elements of the village. Besides by recycling those dilapidated housing materials, new buildings are regenerated.

场地位置及现状 / Location & Background

主要策略 / Main Strategies

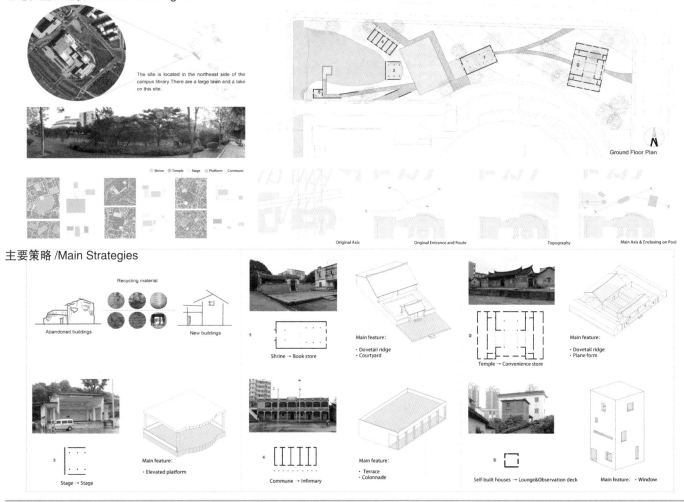

Merging and Symbiosis Conceptual Design for Spatial Sustainable Development of HQU's Xiamen Campus
华侨大学厦门校区校园空间可持续发展概念设计
2016年五校联合毕业设计终期汇报

融合·共生与介入·整合 —— 华侨大学厦门校区与村落的城市设计研究

单体设计

 卢汀滢 Lu Tingying

EMBEDDING 出砖入石

指导老师：张路峰 2016-6-20

B2 - 音舞公园
B2 - Music & Dancing Park

音舞公园坐落在音乐舞蹈学院旁边。在该设计中，我们试图在场地上置入微小但重要的功能，使得村民在学校内也能有自己的领域，他们的子女也可以在音舞公园中进行活动。
The Yinwu Park locates beside the Music And Dancing Institution. In this project, we are trying to embed tiny but vital function blocks on the site, and at the same time make it possible to cross and do activities in the Yinwu Park.

1-1 Section

该场地原本就是绿化用地，所以我们不能建造大体量建筑。
Originally, the site has been designed as a piece of green land, so we are not allowed to construct huge architectures.

功能置入 /Function Inserting

总平面图
Site Plan

图解分析 /Program

第一步 /Step 1

原有的闲置绿地充满着新的可能性，但人们却只能从它旁边绕行。

第二步 /Step 2

为了划分场地使之发挥出多样的可能性，我们利用音舞学院和西侧图书馆建立新道路。

第三步 /Step 3

潜在的路径将场地自然划分成块，成为我们引入新功能的前提。

 Merging and Symbiosis Conceptual Design for Spatial Sustainable Development of HQU's Xiamen Campus
华侨大学厦门校区校园空间可持续发展概念设计
2016年五校联合毕业设计终期汇报

上篇：融合与共生——华侨大学厦门校区校园空间可持续发展概念设计

单体设计 / EMBEDDING 出砖入石

卢汀滢 Lu Tingying　指导老师：张路峰 2016-6-20

S2 - 华侨文化博物馆 / Overseas Chinese Culture Museum

华侨华人有着独特的文化，与海外亲眷有着密切的联系。这个博物馆不仅是为学校服务，也为市民和他们的海外亲眷服务。传统大厝是侨村建筑的重要代表，在本设计中，我们试图找出中国传统民居的要素，并将其用现代的手法重新设计和演绎。

The overseas Chinese has its unique culture which is closely linked with the relatives in foreign countries. This museum is not only for the school, but also for the villagers and their overseas relatives. The traditional form of Da Cuo is a symbol of overseas Chinese villages. In this project, we are trying to find out the identification of Chinese traditional houses and redesign them into a modern form.

1-1 剖透视图 / 1-1 Section Perspective

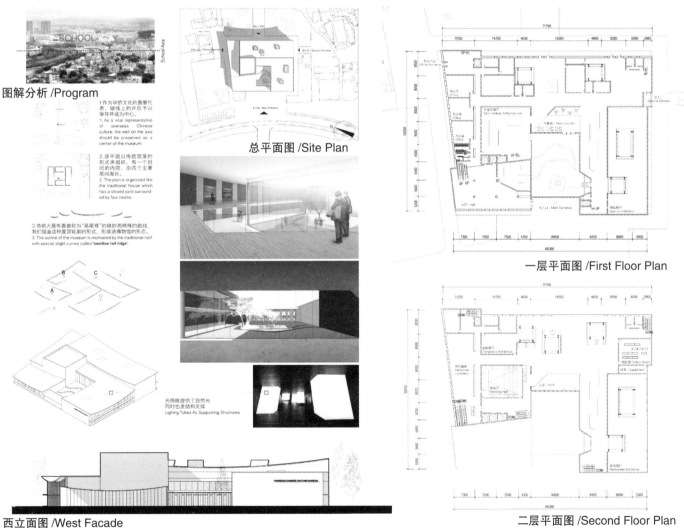

图解分析 / Program

总平面图 / Site Plan

一层平面图 / First Floor Plan

二层平面图 / Second Floor Plan

西立面图 / West Facade

Merging and Symbiosis Conceptual Design for Spatial Sustainable Development of HQU's Xiamen Campus
华侨大学厦门校区校园空间可持续发展概念设计
2016年五校联合毕业设计终期汇报

UCAS

融合·共生与介入·整合 —— 华侨大学厦门校区与村落的城市设计研究

化整为零
PARTED WHOLE

指导老师：张路峰 课程助教：彭相国
设计者：李偲淼 丁沫 彭宁

PARTED WHOLE 化整爲零

李偲淼 Li Simiao　丁沫 Ding Mo　彭宁 Peng Ning

指导老师：张路峰　2016-6-20

问题分析与界定

建筑现状

古厝 Old Building
家庙与宗祠 Temple and Ancestor

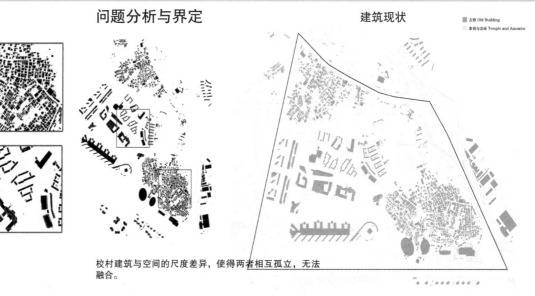

校村建筑与空间的尺度差异，使得两者相互孤立，无法融合。

校园功能分区

教学区 Teaching Area
生活区 Living Area
文博区 Museum Area
运动区 Sports Area
自然区 Nature Area

开放绿地

农田 Farmland
广场 Square
水塘 Pond
绿化 atforest
空地 Vacant Place

绿地与公共空间分析

村落分析

北部村落

南部村落

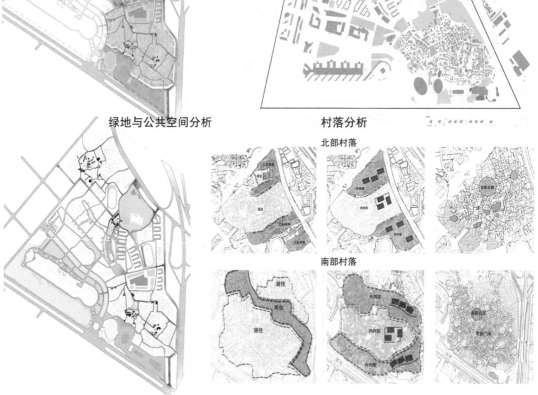

教师评语：

该作品聚焦校园空间扩张背景下的校、村空间在尺度上的冲突。新建校园的大尺度建筑要满足学校每项活动的需要，而大量建筑的占地会导致对原有村落的破坏，进而有损于原有的社会关系和华侨文化。作品巧妙地将大尺度公共建筑的各类功能空间加以拆解，并分散植入原村落。一片片传统建筑群在保持原空间秩序的同时，在功能上也满足了学校的需要。

01　Merging and Symbiosis Conceptual Design for Spatial Sustainable Development of HQU's Xiamen Campus
华侨大学厦门校区校园空间可持续发展概念设计
2016年五校联合毕业设计终期汇报

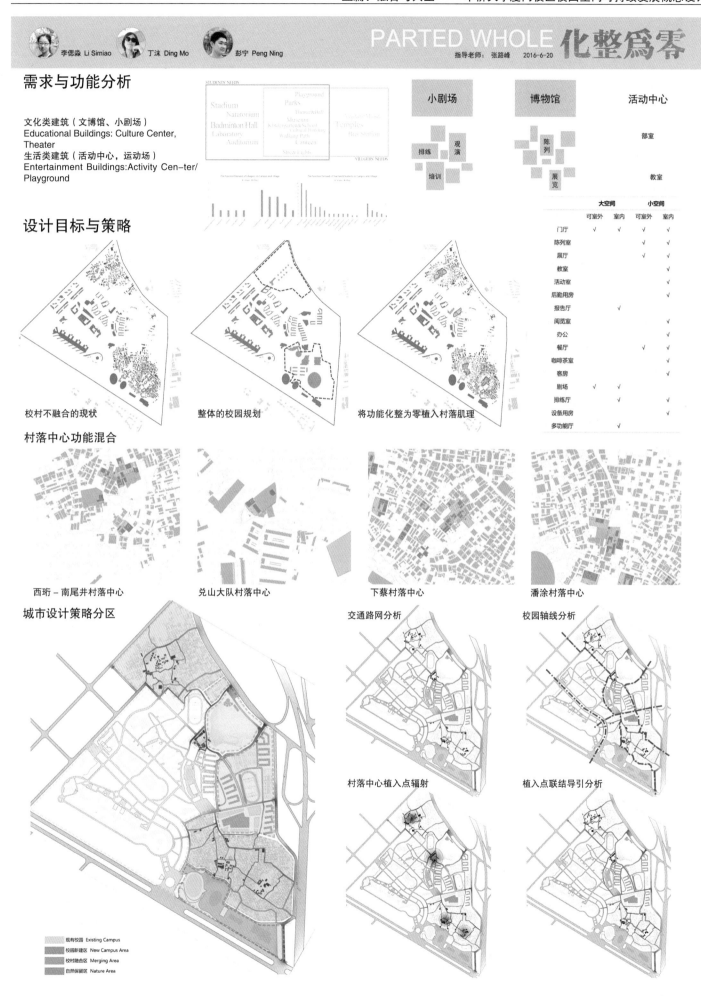

融合·共生与介入·整合 —— 华侨大学厦门校区与村落的城市设计研究

PARTED WHOLE 化整爲零

李偲淼 Li Simiao　丁沫 Ding Mo　彭宁 Peng Ning

指导老师：张路峰　2016-6-20

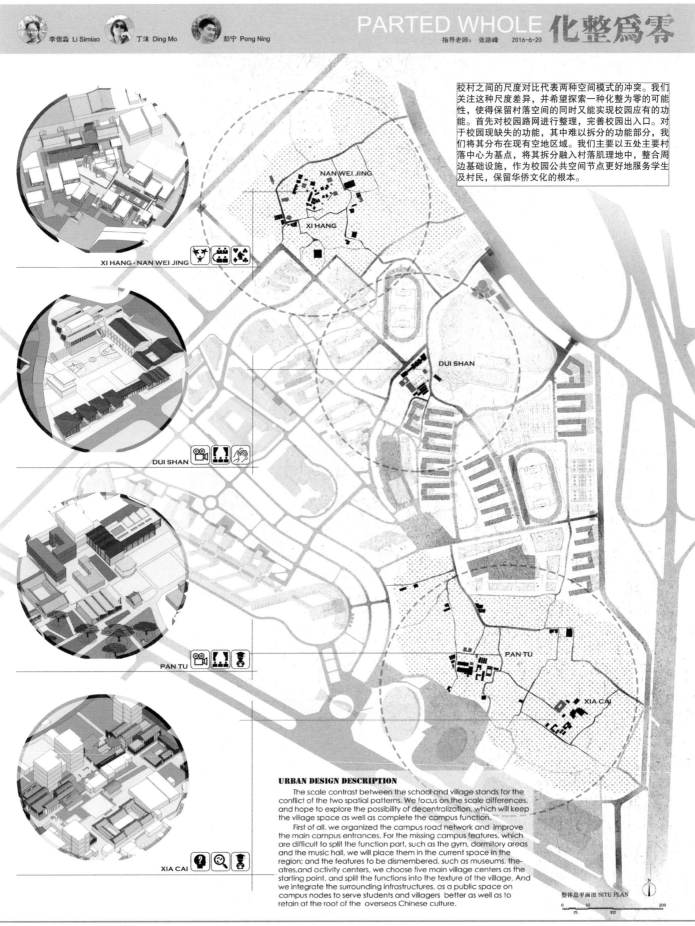

校村之间的尺度对比代表两种空间模式的冲突。我们关注这种尺度差异，并希望探索一种化整为零的可能性，使得保留村落空间的同时又能实现校园应有的功能。首先对校园路网进行整理，完善校园出入口。对于校园现缺失的功能，其中难以拆分的功能部分，我们将其分布在现有空地区域。我们主要以五处主要村落中心为基点，将其拆分融入村落肌理地中，整合周边基础设施，作为校园公共空间节点更好地服务学生及村民，保留华侨文化的根本。

URBAN DESIGN DESCRIPTION

The scale contrast between the school and village stands for the conflict of the two spatial patterns. We focus on the scale differences, and hope to explore the possibility of decentralization, which will keep the village space as well as complete the campus function.

First of all, we organized the campus road network and improve the main campus entrances. For the missing campus features, which are difficult to split the function part, such as the gym, dormitory areas and the music hall, we will place them in the current space in the region; and the features to be dismembered, such as museums, theatres, and activity centers, we choose five main village centers as the starting point, and split the functions into the texture of the village. And we integrate the surrounding infrastructures, as a public space on campus nodes to serve students and villagers better as well as to retain at the root of the overseas Chinese culture.

整体总平面图 SITE PLAN

Merging and Symbiosis Conceptual Design for Spatial Sustainable Development of HQU's Xiamen Campus

华侨大学厦门校区校园空间可持续发展概念设计
2016年五校联合毕业设计终期汇报

上篇：融合与共生 —— 华侨大学厦门校区校园空间可持续发展概念设计

单体设计　李偲淼 Li Simiao　　PARTED WHOLE　化整爲零
指导老师：张路峰　2016-6-20

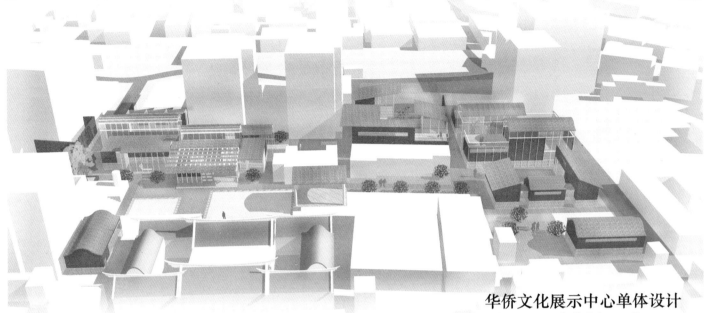

华侨文化展示中心单体设计
Overseas Chinese Culture Exhibition Center

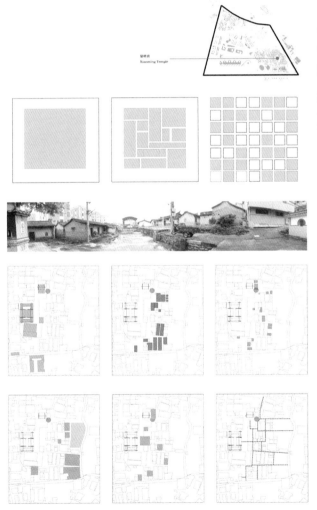

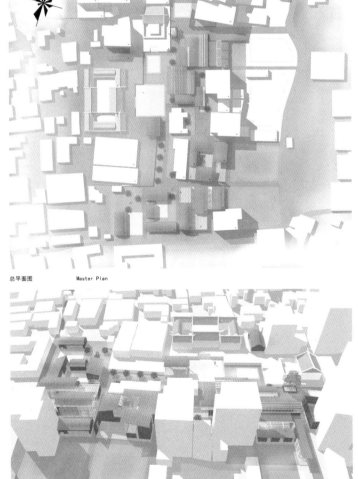

总平面图　Master Plan

Merging and Symbiosis Conceptual Design for Spatial Sustainable Development of HQU's Xiamen Campus
华侨大学厦门校区校园空间可持续发展概念设计
2016年五校联合毕业设计终期汇报

融合·共生与介入·整合 —— 华侨大学厦门校区与村落的城市设计研究

单体设计　PARTED WHOLE　化整爲零

李偲淼 Li Simiao　　指导老师：张路峰　2016-6-20

一层平面图 First Floor Plan

1. 华侨文化展厅
2. 小型文化图书馆
3. 图书馆阅读室
4. 图书馆研习室
5. 公共活动室
6. 自助水吧
7. 休憩室
8. 活动展示厅
9. 公共教育沙龙
10. 沙龙讨论区
11. 工作坊接待处
12. 文化创意工作坊
13. 文化咖啡厅
14. 物品贩售

1. Culture Gallery
2. Small Library
3. Reading Room
4. Discussion Room
5. Public Activity
6. Pantry
7. Rest Room
8. Activity Display
9. Education Salon
10. Discussion Area
11. Studio Acception
12. Creation Studio
13. Cafe
14. Sales

二层平面图 Second Floor Plan

1. 活动室　　1. Activity Room
2. 办公室　　2. Manage Office
3. 景观台　　3. View Balcony
4. 咖啡厅　　4. Cafe(2 floor)
5. 工作室　　5. Culture Studio
6. 屋顶花园　6. Roof Garden

重构策略 Reconstructed Strategy

改造的建筑 Reconstructed Buildings

古建筑墙体分布 Walls of Old Buildings

拆除部分墙体进行空间重组 Space Organize by Demolishing Walls

剖视图 Perspective Sections

商品贩售 Sales
文创工作室 Workshop
咖啡厅 Cafe

景观廊道 View Corridor
休息区 Restroom
茶水吧 Tearoom
活动区 Public Activity

研习室 Discussion Room
文化资料馆 Culture Library
文化展廊 Exhibition Corridor
文化展厅 Exhibition Hall
景观廊道 View Corridor

立面图 Elevation

Merging and Symbiosis Conceptual Design for Spatial Sustainable Development of HQU's Xiamen Campus
华侨大学厦门校区校园空间可持续发展概念设计
2016年五校联合毕业设计终期汇报

上篇：融合与共生 —— 华侨大学厦门校区校园空间可持续发展概念设计

单体设计 彭宁 Peng Ning

PARTED WHOLE 化整为零

指导老师：张路峰　2016-6-20

文化剧场设计
Design of Culture Theatres

设计说明：
　　除音乐厅外，校村双方对于文艺演出和观演的功能也存在需求。按照城市设计"化整为零"的规则，除大空间的音乐厅外，诸如小剧场、排练厅、活动室等功能空间都是小尺度且功能并置的，因此可以分散在不同的村落中心，实现化整归零。

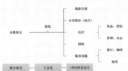

根据不同区位的现状条件和城市设计规则，文化剧场体系的设计将采取不同的手法。
1. 兑山大队村落中心
新建小型剧场和活动室等，完善村落肌理。
2. 校园可建设空地
新建大型音乐厅，结合道路形成完整的校园轴线。
3. 潘涂村落中心
改建周边废弃或破损房屋，作为活动室或琴房。
4. 下蔡、西珩等村落中心
翻修戏台，重塑戏台周边的室外观演空间。

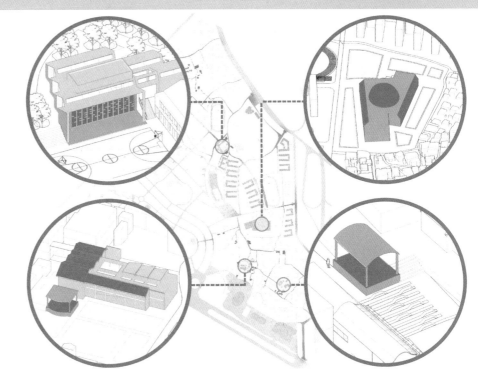

潘涂村落设计
Pantu Village Center

总平面图　Site Plan

1 室外场地 Outdoor Ground
2 底层架空 Bottom Overhead
3 排练室 Rehearsal Room
4 活动室 Activity Room
5 办公 Office

一层平面图　First Floor Plan

1 二层大厅 Second Floor Hall
2 练琴房 Practice Room
3 活动室 Activity Room
4 室外平台 Outdoor Terrace

二层平面图　Second Floor Plan

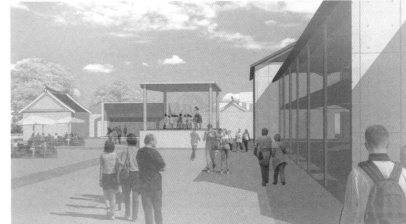

立面图 Elevation

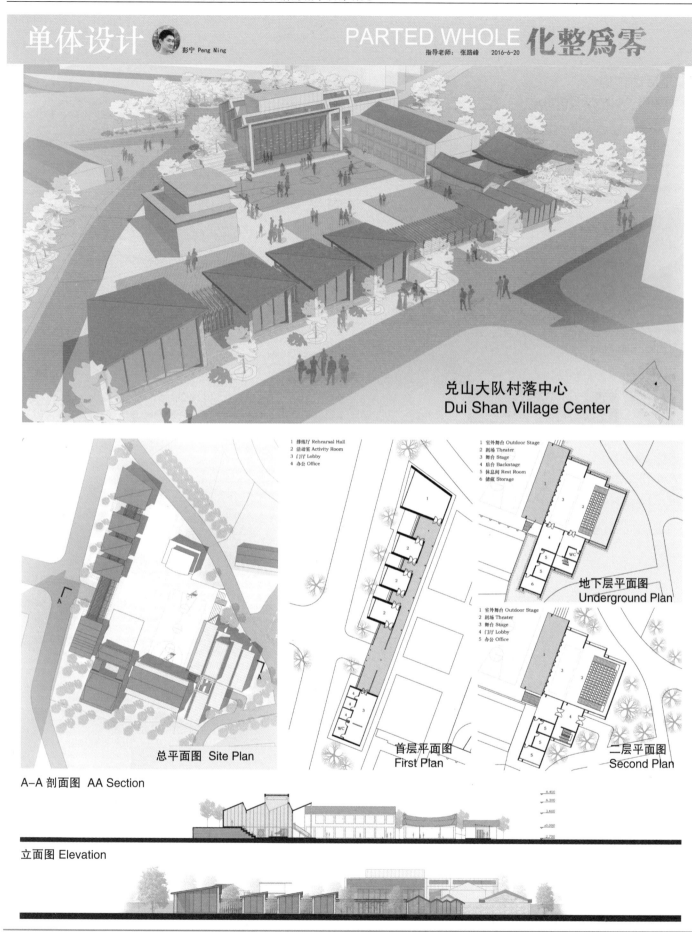

PARTED WHOLE 化整為零

单体设计 — 彭宁 Peng Ning
指导老师：张路峰　2016-6-20

兑山大队村落中心　Dui Shan Village Center

总平面图　Site Plan

地下层平面图　Underground Plan

首层平面图　First Plan

二层平面图　Second Plan

A-A 剖面图　AA Section

立面图　Elevation

Merging and Symbiosis Conceptual Design for Spatial Sustainable Development of HQU's Xiamen Campus
华侨大学厦门校区校园空间可持续发展概念设计
2016年五校联合毕业设计终期汇报

单体设计 / PARTED WHOLE / 化整爲零

丁沫 Ding Mo　　指导老师：张路峰　2016-6-20

学生活动中心 / Student Activity Center

设计说明：
将大学生活动中心分散到多个村落中心，服务于周边可达的学生与居民，从而更高效地辐射校园边缘区域。活动中心位于两个中心之间，整合原有道路与空地，形成一条联系通道，串联多个重要因素。

总平面图　Site Plan

透视图　Perspective Plan

二层平面图　Second Floor Plan

一层平面图　First Floor Plan

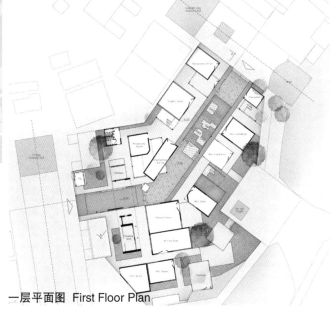

东立面图　East Elevation

南立面图　South Elevation

华侨大学
HUAQIAO UNIVERSITY

目录 CATALOG

公共服务导向的校村融合与共生　方弘毅　王晓宁　杨成颢
The Public Service Oriented

水文复兴　　　　　　　　　　　林丰　王磊　何吟涛
The Renaissance of Water

差异的织补　　　　　　　　　　李泰　彭永宸
Weave the Conflict

学生团队 Student Team

李泰

彭永宸

林丰

王磊

何吟涛

方弘毅

杨成颢

王晓宁

WEAVE THE CONFLICT
差异的织补——华侨大学厦门校区校园空间可持续发展概念设计

差异的织补 WEAVE THE CONFLICT

指导老师：刘塨 郑志
课程助教：姚敏峰 赖世贤
设计者：李泰 彭永宸

ISOMETRIC DRAWING

- LI TEMPLE
- ANCESTRAL TEMPLE
- DUISHAN FOREST PARK
- TRADITIONAL ARCHITECTURE
- GURONG PARK
- OLD TREES
- GREEN

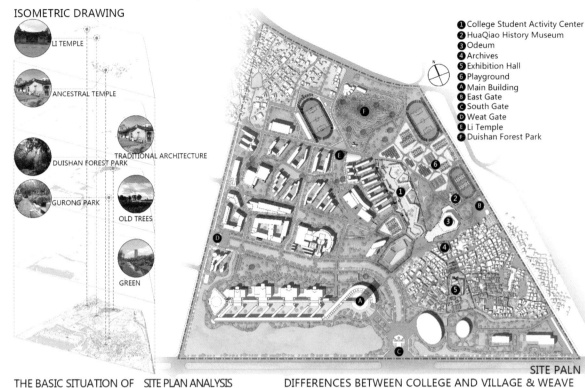

1. College Student Activity Center
2. HuaQiao History Museum
3. Odeum
4. Archives
5. Exhibition Hall
6. Playground
A. Main Building
B. East Gate
C. South Gate
D. Weat Gate
E. Li Temple
F. Duishan Forest Park

SITE PALN

THE BASIC SITUATION OF THE COLLEGE & VILLAGE

VILLAGE DISTRIBUTION (XIHANG, NANWEIJING, XITOU, ZUCUOHOU, PANTU, ZHAINEI, XIACAI)

BASE (College, Village)

COLLEGE (Living area, Teaching area)

VILLAGE (Village, Park/Tomb, Water company/Launch tower, Village)

SITE PLAN ANALYSIS

ROAD

GREEN&WATER

AXIS

ACTIVITY PLACE

DIFFERENCES BETWEEN COLLEGE AND VILLAGE & WEAVE

1. WEAVE ROAD

COLLEGE: Road network scale is large | VILLAGE: Road network scale is small | Optimization of road and weave them

2. WEAVE ACTIVITY PLACE

Weaving network and sharing site | VILLAGE: A lot of free space | COLLEGE: Lack of playground

3. WEAVE ACTIVITIES

COLLEGE | VILLAGE | Add new activities

4. WEAVE HUMAN COMMUNICATION

Students | Villagers | More communication

教师评语：

突出的特点是通过中心与边界关系的几何学形态拓扑变换，把原有地的各种元素之间的间隙挤压挤之间的差异得到织补。把城市设计范畴内几乎不可解的难题，转变为拓扑几何学的问题，使之有了理性的求解过程与方法。这是一种非常具有启发性的转换。这种转换中呈现的对称性，具有深刻的人居环境的意义。作品也因而呈现不寻常的结果。

非常希望这个作业的方法与结果，通过进一步的深入探索确认普遍意义。

WEAVE THE CONFLICT
差异的织补——华侨大学厦门校区校园空间可持续发展概念设计

DIAGRAM OF CONCEPT

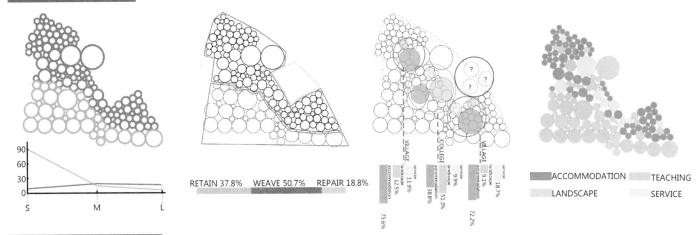

TIME AND MOLDING SYSTEM

ELEMENT: PLAYGROUND=5 ACTIVITY+ACCOMMODATION=20 GREEN=25   DRAG=10 ATTRACTION=14 SPRING=6

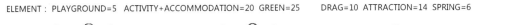

EMERGENCE OF RANDOM RESULTS

MESH GENERATION

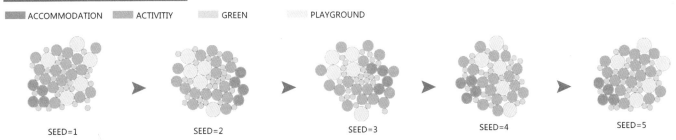

VILLAGE INTEGRATION

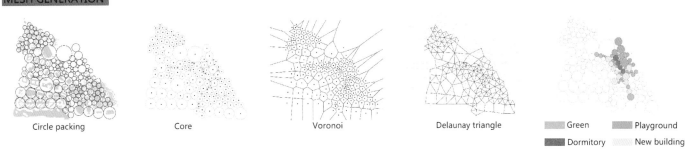

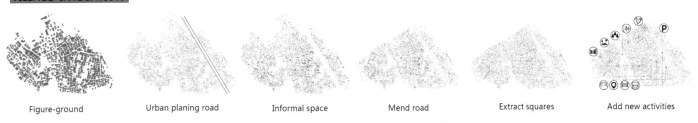

WEAVE THE CONFLICT
差异的织补——华侨大学厦门校区校园空间可持续发展概念设计

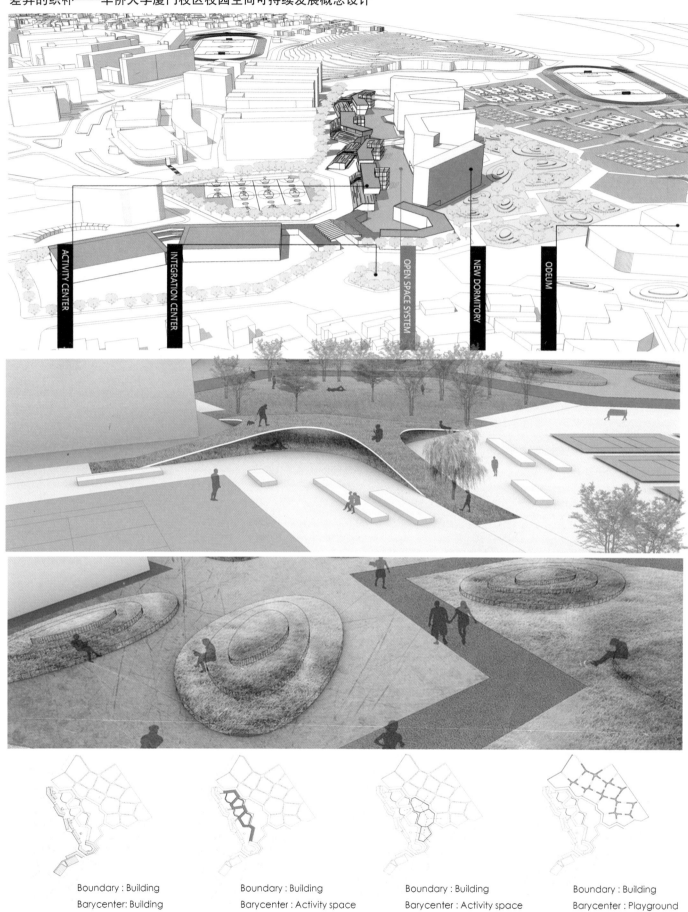

| Boundary : Building | Boundary : Building | Boundary : Building | Boundary : Building |
| Barycenter: Building | Barycenter : Activity space | Barycenter : Activity space | Barycenter : Playground |

WEAVE THE CONFLICT
差异的织补——华侨大学厦门校区校园空间可持续发展概念设计

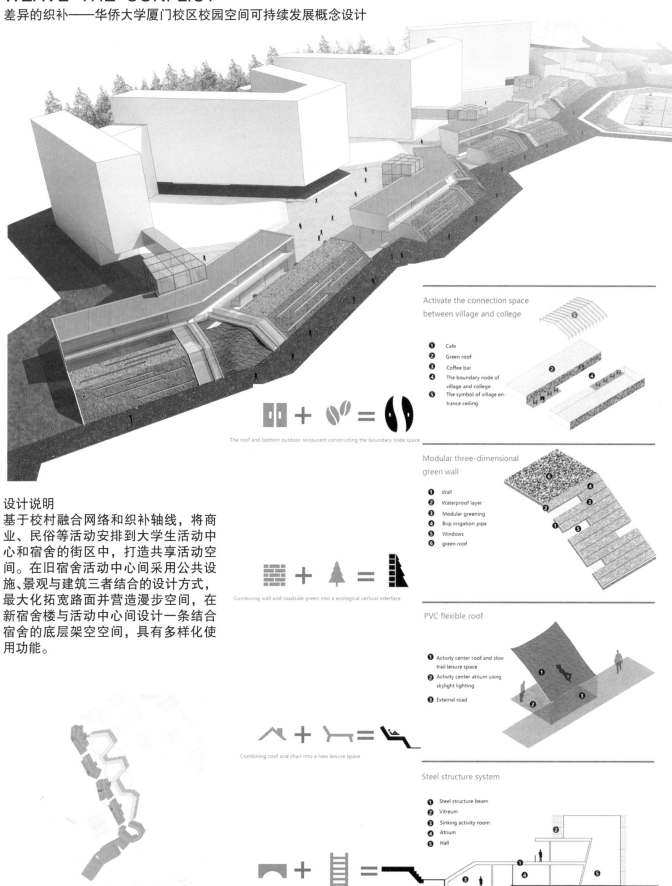

设计说明

基于校村融合网络和织补轴线，将商业、民俗等活动安排到大学生活动中心和宿舍的街区中，打造共享活动空间。在旧宿舍活动中心间采用公共设施、景观与建筑三者结合的设计方式，最大化拓宽路面并营造漫步空间，在新宿舍楼与活动中心间设计一条结合宿舍的底层架空空间，具有多样化使用功能。

融合·共生与介入·整合——华侨大学厦门校区与村落的城市设计研究

WEAVE THE CONFLICT
差异的织补——华侨大学厦门校区校园空间可持续发展概念设计

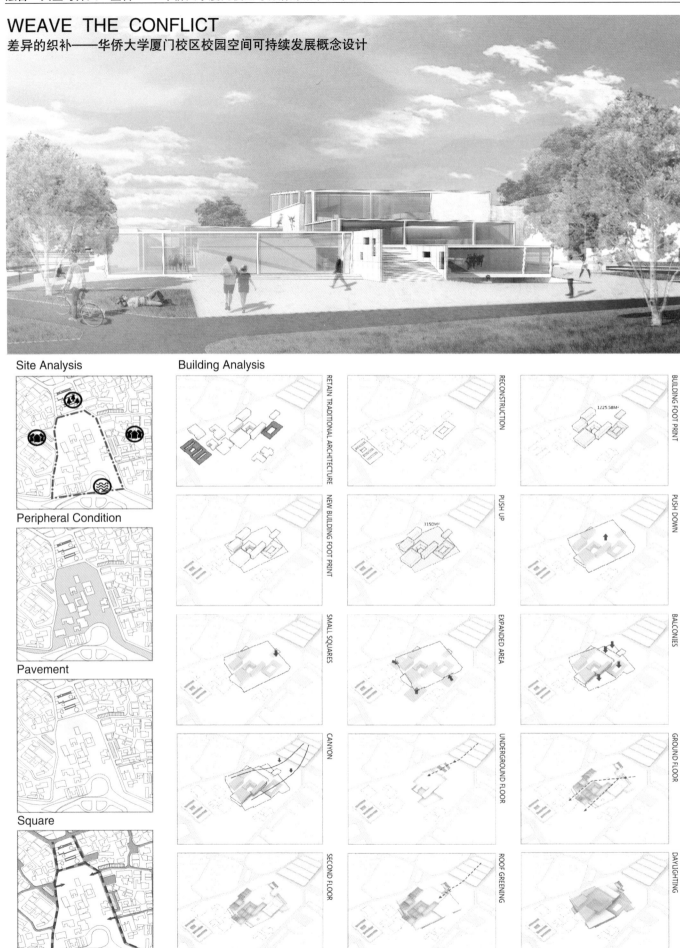

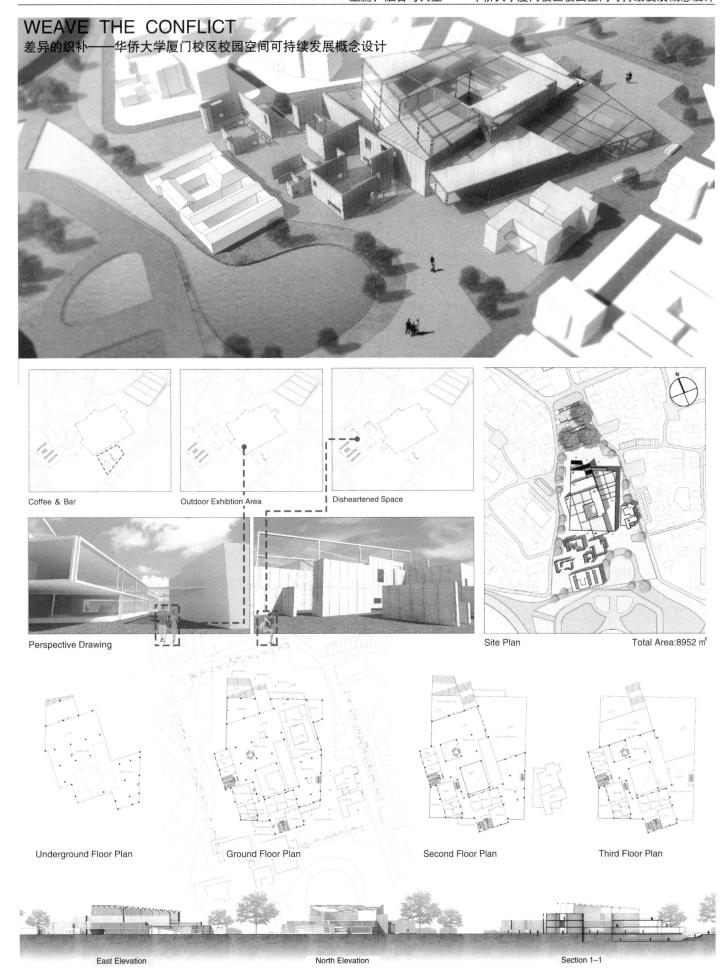

融合·共生与介入·整合——华侨大学厦门校区与村落的城市设计研究

水文复兴
The Renaissance of Water
——融合共生 五校联合设计

指导老师：刘塨 郑志　课程助教：姚敏峰 赖世贤
设计者：何吟涛 林丰 王磊

教师评语：

作品注意到场地中水系对于环境的生态学与形态结构的双重整合意义，通过把校园所依托的水系湿地功能逻辑向村落延展，并与村落中的水系元素有机融合，从而达成"校-村"机能与形态结合生态学结构的双重融合。在两者的融合中，村落提升了环境生态与社会学价值，学校也得到来自村落之活力的补充。

现状水分析 WATER SITUATION ANALYSIS

水系 Hydro-System

水流来向 Water Input

水流去向 Water Output

绿化关系 Green Relation

车流关系 Ketchup

人流关系 Pedesttian

现状问题 CURRENT SITUATION

池塘 POND
南尾井村中心前有保留池塘，但是水流不畅，经常发出恶臭。

污水明沟 SEWAGE DITCH
村庄污水经常通过明沟排出，导致村庄在视觉和味觉上体验不佳。

污水 SEWAGE
部分污水直接排入校园湿地，给湿地系统造成极大的压力。

污水暗沟 SEWAGE DITCH
暗沟是理想的排污方式，但是村中暗沟较少。

池塘 POND
下蔡村中的池塘已经长满绿藻，环境恶劣。

填埋池塘 BURIED POND
近些年村民将池塘填埋种植蔬菜湿地系统破坏严重。

池塘 Pond
保护村落原有水塘，形成具有场所精神的空间。也为两边学生宿舍提供良好的景观。

荷花池 Lotus Pond
亦为村落原有水塘，因建设图书馆被填埋，后重新恢复，种植荷花。

功能湿地 Function Wetland
由西门连接至主楼的功能湿地是连通校园水系的主要廊道，在生态学和校园景观上有重要作用。

斑块 Plaque
由于地势较低，易于汇水。将此设计成为斑块，为动物栖息繁殖提供场所也调节校园内的微气候。

中水厂 Reclaimed Water
回收一万两千名学生的生活污水重新利用。最程度保护环境。

校门 School Gate
校门的设计也使用水作为意向，半圆形拱门在水的映衬下变成完整的圆环。

历史比较 COMPARATIVE HISTORY

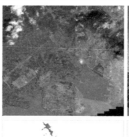
村落与水系历史分布图
Water History in Duisham

村落与水系现状分布图
Water Present in Duisham

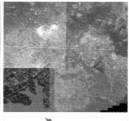

2005年

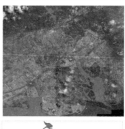

2010年　2015年

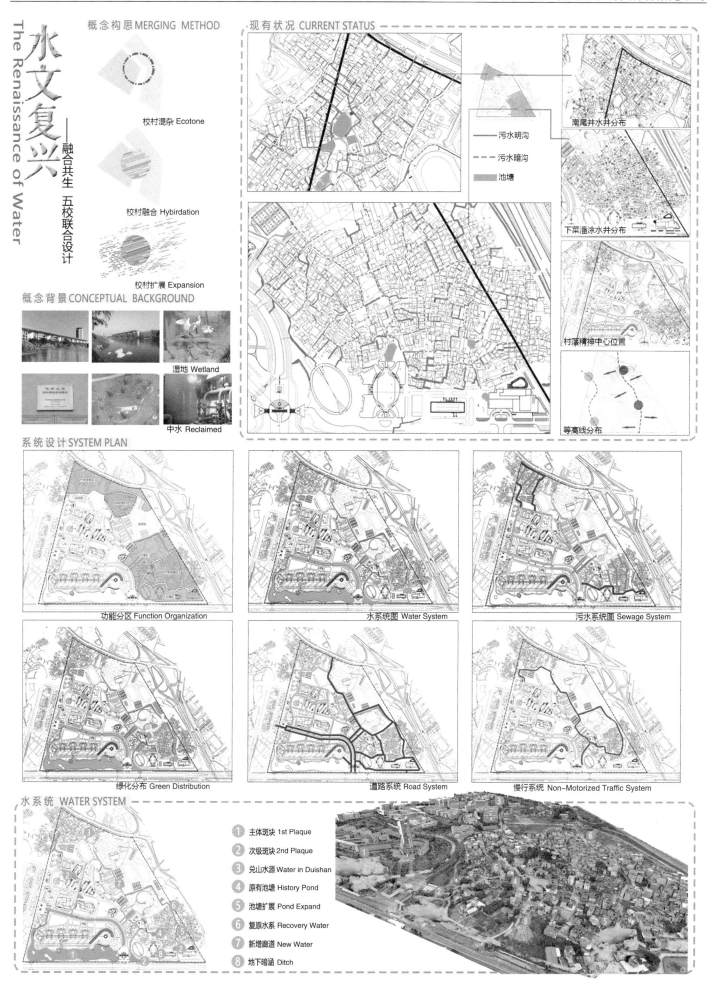

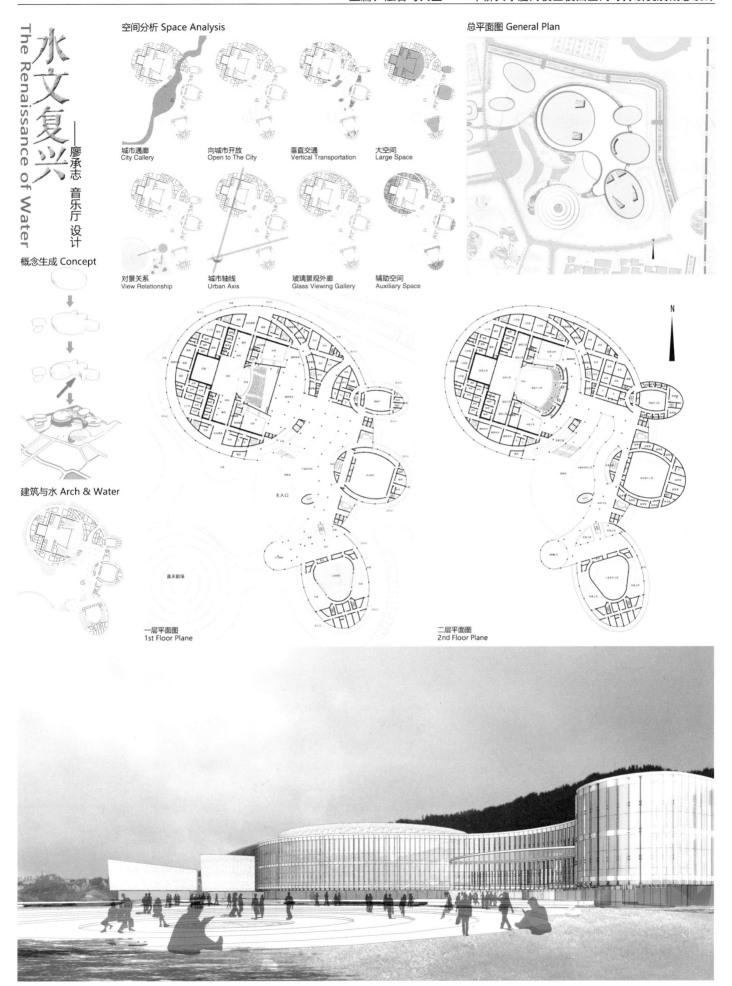

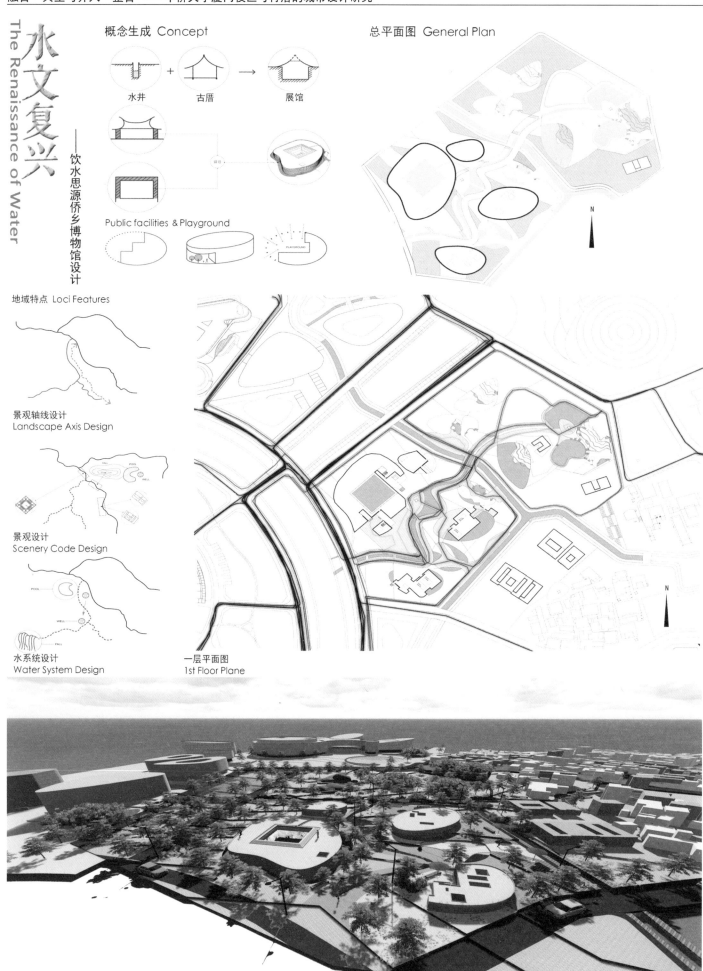

上篇：融合与共生 —— 华侨大学厦门校区校园空间可持续发展概念设计

水文复兴 The Renaissance of Water
——大学生活动中心设计

设计说明 Specification

大学生活动中心根据场地轴线和水系进行设计，将音乐舞蹈学院与音乐厅之间的轴线进行串联，建筑划分为四个组团，建筑外部空间延续城市设计的思路，吸引村民和学生共同参与，促进融合。

The center of the student activity center is designed according to the site axis and the water system. The building is divided into four groups according to the axis between the music and dance school and the concert hall. Building external space to continue the idea of urban design, to attract villagers and students to participate in, to promote integration.

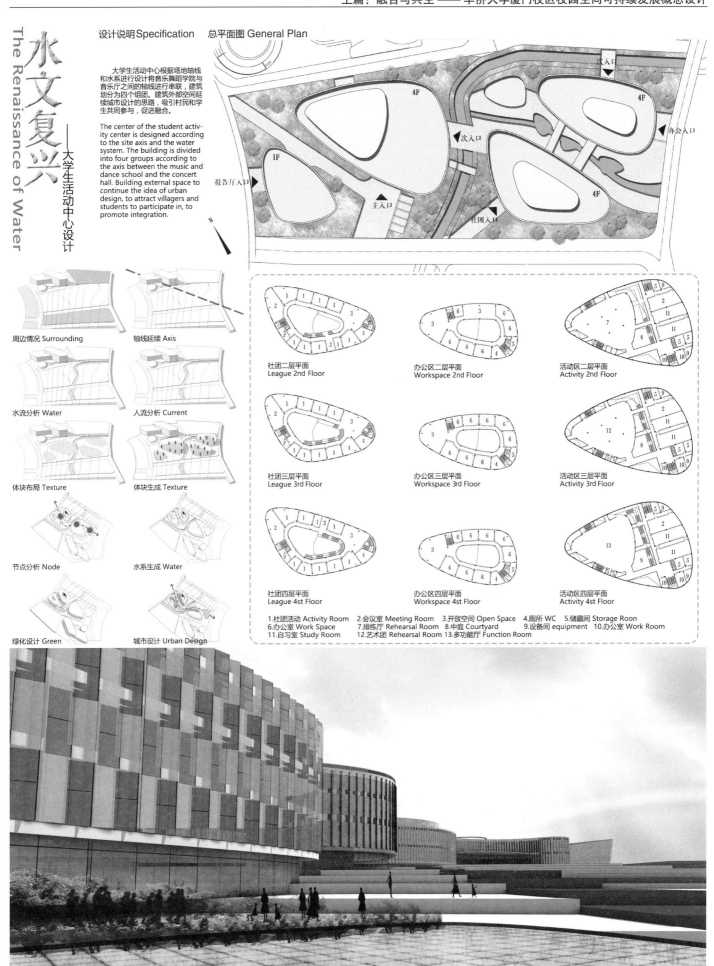

总平面图 General Plan

周边情况 Surrounding　轴线延续 Axis
水流分析 Water　人流分析 Current
体块布局 Texture　体块生成 Texture
节点分析 Node　水系生成 Water
绿化设计 Green　城市设计 Urban Design

社团二层平面 League 2nd Floor　办公区二层平面 Workspace 2nd Floor　活动区二层平面 Activity 2nd Floor
社团三层平面 League 3rd Floor　办公区三层平面 Workspace 3rd Floor　活动区三层平面 Activity 3rd Floor
社团四层平面 League 4st Floor　办公区四层平面 Workspace 4st Floor　活动区四层平面 Activity 4st Floor

1.社团活动 Activity Room　2.会议室 Meeting Room　3.开放空间 Open Space　4.厕所 WC　5.储藏间 Storage Roon
6.办公室 Work Space　7.排练厅 Rehearsal Room　8.中庭 Courtyard　9.设备间 equipment　10.办公室 Work Room
11.自习室 Study Room　12.艺术团 Rehearsal Room　13.多功能厅 Function Room

融合·共生与介入·整合——华侨大学厦门校区与村落的城市设计研究

The Public Service Oriented
Conceptual Design for Spatial Sustainable Development of HQU's Xiamen Campus

公共服务导向的校村融合与共生
THE PUBLIC SERVICE ORIENTED

指导老师：刘塨 郑志　课程助教：姚敏峰 赖世贤
设计者：方弘毅 王晓宁 杨成颢

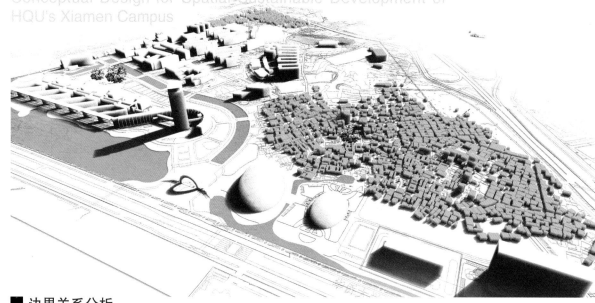

■ 边界关系分析

| 基地分布 /Base | 村落场地 /Village | 村落分布 /Village Distribution | 校村交界功能 /Frontier |

学校 /College
村庄 /Village

村庄 /Village
公园 /Park 陵墓 /Lamb
水务集团 /Water Company

大门 /Gate
宿舍 /Dormitory
食堂 /Restaurant
华文学院 /Huawen Academy

■ 场地分析

建筑 /Architecture　　公共服务 /Public Service　　交通 /Traffic　　中心空间 /Centre

现代建筑 /Architecture
传统建筑 /Traditional Architecture

超市 /Market　工厂 /Factory
幼儿园 /Kindergarten

主干道 /High streets
次干道 /Collector streets

入口节点 /Node

建筑 /Architecture　　公共服务 /Public service　　交通 /Traffic　　中心空间 /Centre

    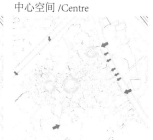 

现代建筑 /Architecture
传统建筑 /Traditional Architecture

超市 /Market　工厂 /Factory
幼儿园 /Kindergarten

主干道 /High Streets
次干道 /Collector Streets

入口节点 /Node

教师评语：

作品发现了校园与村落在公共服务领域的需求与发展具有高度的同一性，以此作为两者融合发展的动力与思路。作品进而将校园秩序延伸叠加于村落肌理之上，有次序性地放开肌理，村落原有肌理也向校园延伸，这两种方式以形成校村同构以进行交流对话，而公共服务的加入，使这种空间性的融合产生了社会学中最具活力的含义。

上篇：融合与共生 —— 华侨大学厦门校区校园空间可持续发展概念设计

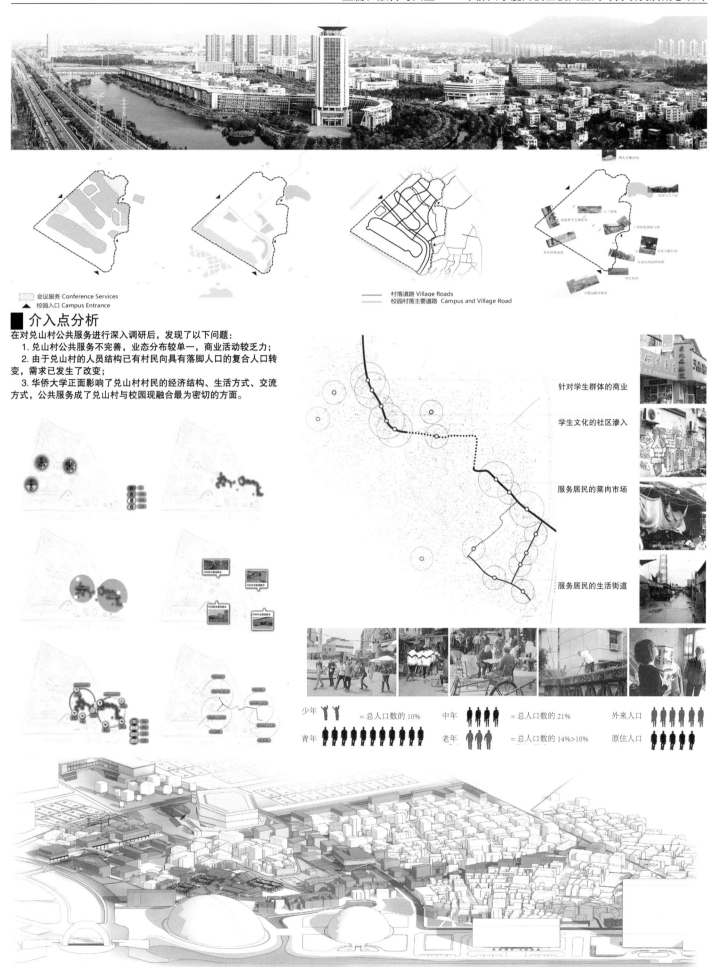

介入点分析

在对兑山村公共服务进行深入调研后，发现了以下问题：
1. 兑山村公共服务不完善，业态分布较单一，商业活动较乏力；
2. 由于兑山村的人员结构已由村民向具有落脚人口的复合人口转变，需求已发生了改变；
3. 华侨大学正面影响了兑山村村民的经济结构、生活方式、交流方式，公共服务成了兑山村与校园融合最为密切的方面。

融合·共生与介入·整合 —— 华侨大学厦门校区与村落的城市设计研究

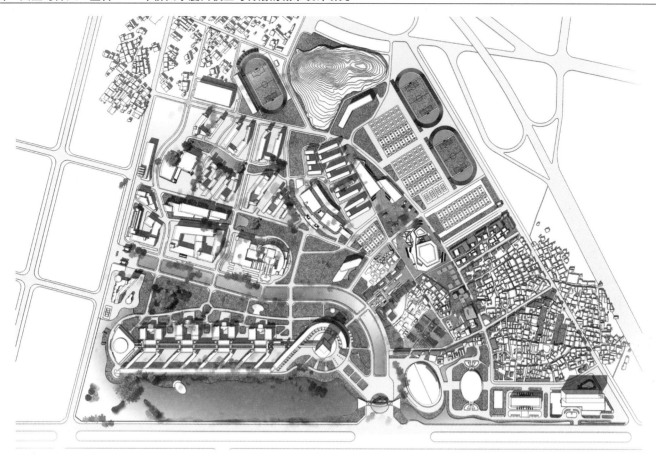

街道重构

Using a variety of methods, combined with the strategy to realize people communicate in business on Main Street, cafe, academic English corner and activity under the tree.

The corner of the street parks, parking location exchange. Let the public street atmosphere thicker and safer.

商业空间重塑

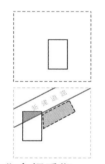

多层次的空间为观赏提供更多的可能性

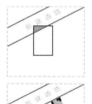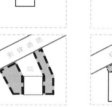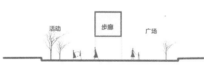

步廊架空与广场上,供行人边走边观赏不同空间的活动

Adding the green and convenient bike lanes. To improve the popularity of the streets, and the essence is human-oriented design.

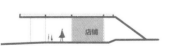

沿街休息场所,为逛街的人群提供休息的场所,更有利于商业的发展

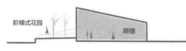

阶梯式花园给停留的人更丰富的空间体验

Guiding people and the car diversion, the closure of a part of function of the street, forming a pocket park. Guiding activities of the villagers and teachers and students in the common public places.

内向性的庭院提供更安详的休息氛围

延续原有上居下商的布置,结合公共服务主轴的广场,发展商业的同时与生活环境相结合,增加商业驻留性

Make good use of the triangle square, after the transformation to attract people to the street, the intergration of public service system interface, the public service node into the Campus-city blocks.

上篇：融合与共生——华侨大学厦门校区校园空间可持续发展概念设计

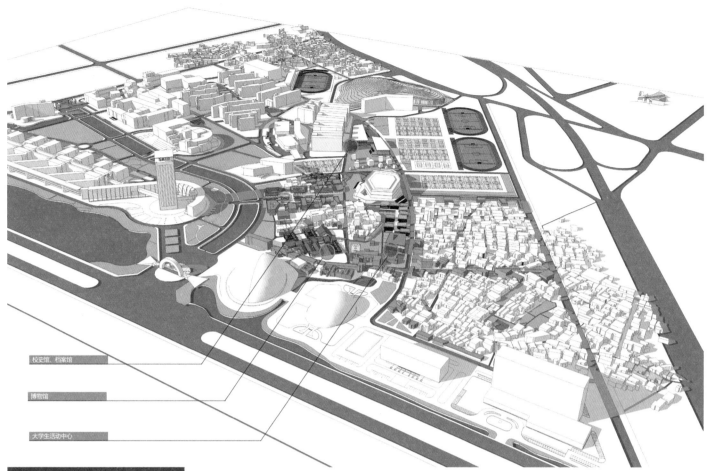

校史馆、档案馆

博物馆

大学生活动中心

■ 街道整合策略

1. 在公共服务网络基础上以学校网络尺度对其进行规划，并与原有村落网络叠加，两套网络共同叠加，从而形成校村共生的网络体系。

2. 新的网络将规整原有公共服务的发展方式，使村落形态的发展更加有序。

这种只整合街面的方式给了原有空间生存的机会，通过时间的发展，新建筑将与旧街区慢慢地融合，城市原有的文化不会消失，最终新添入的元素也会随之沉淀下来，成为城市底蕴的一部分。

■ 重构场所空间 Reconstruction of Space

完全开放空间 适合多种人群 ｜ "L"形广场空间具有场所感 ｜ 半开放空间 ｜ 空间半围合私密性较强

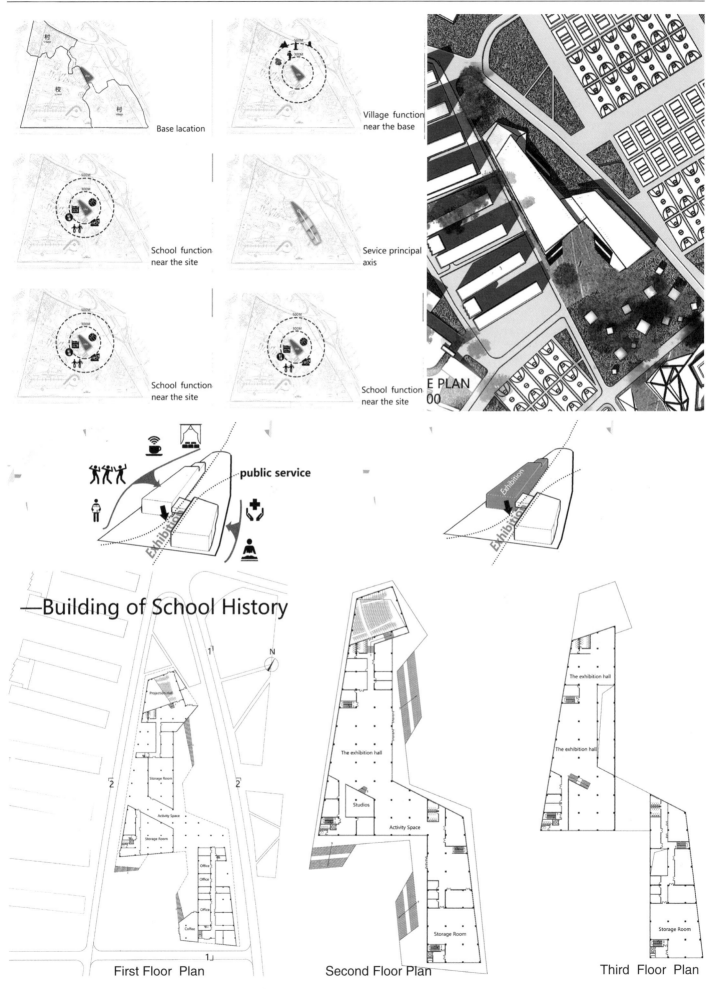

Design of School Archives Center

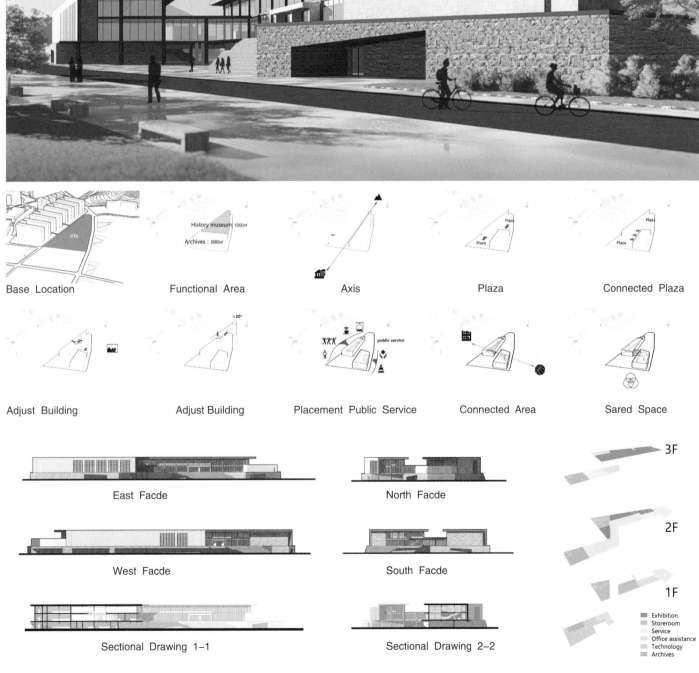

Design of School History Museum

节点透视

节点透视

节点透视

节点透视

First Plan

上篇：融合与共生 —— 华侨大学厦门校区校园空间可持续发展概念设计

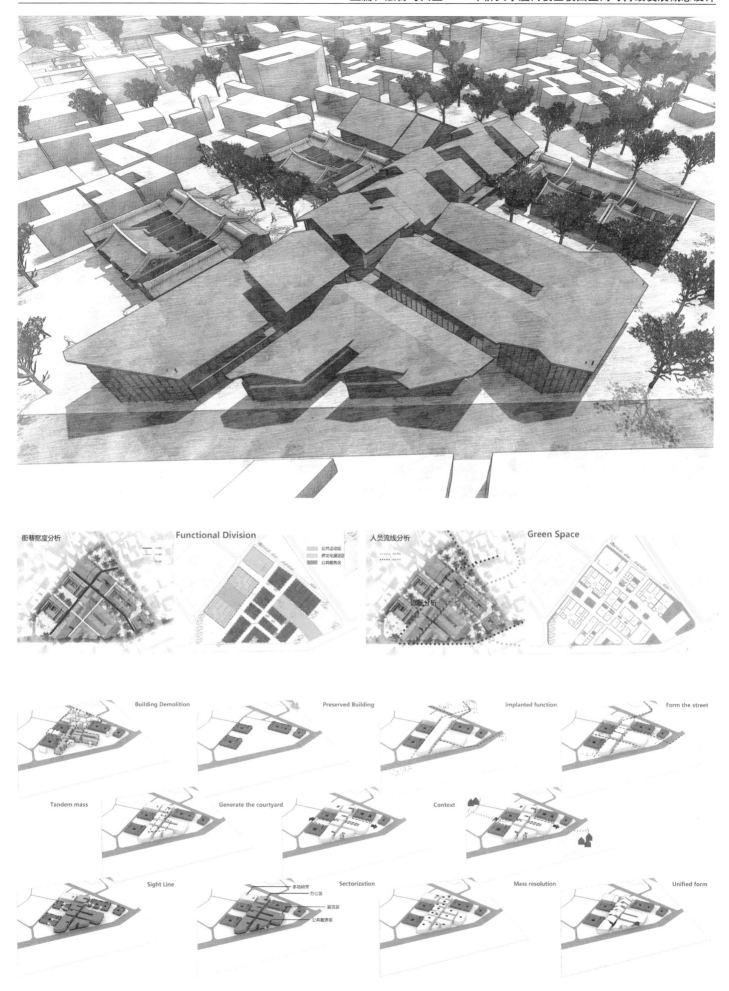

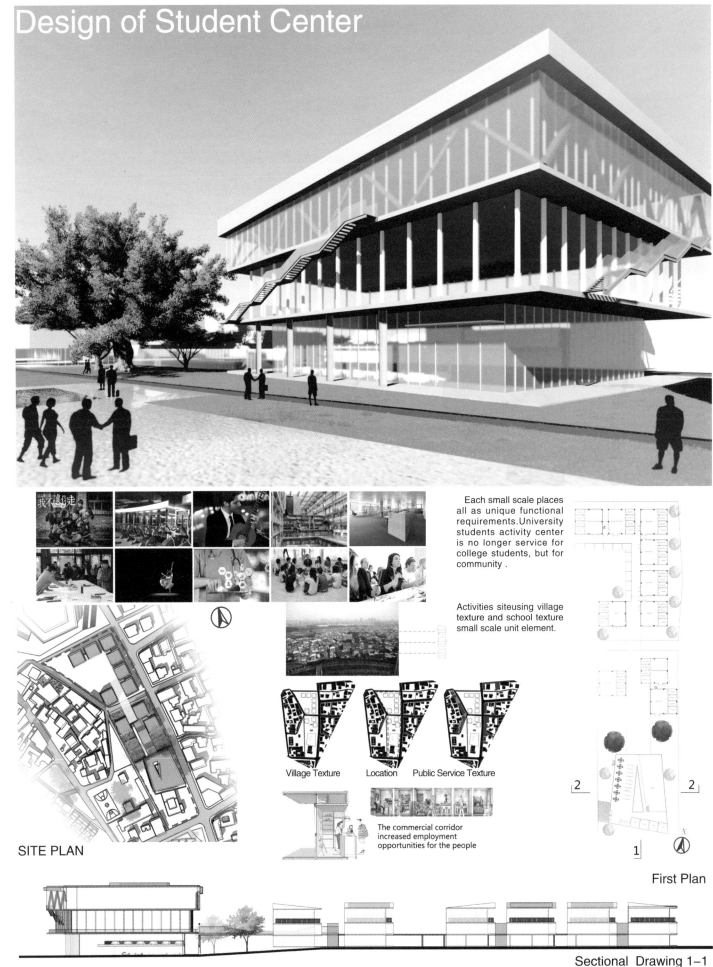

上篇：融合与共生 —— 华侨大学厦门校区校园空间可持续发展概念设计

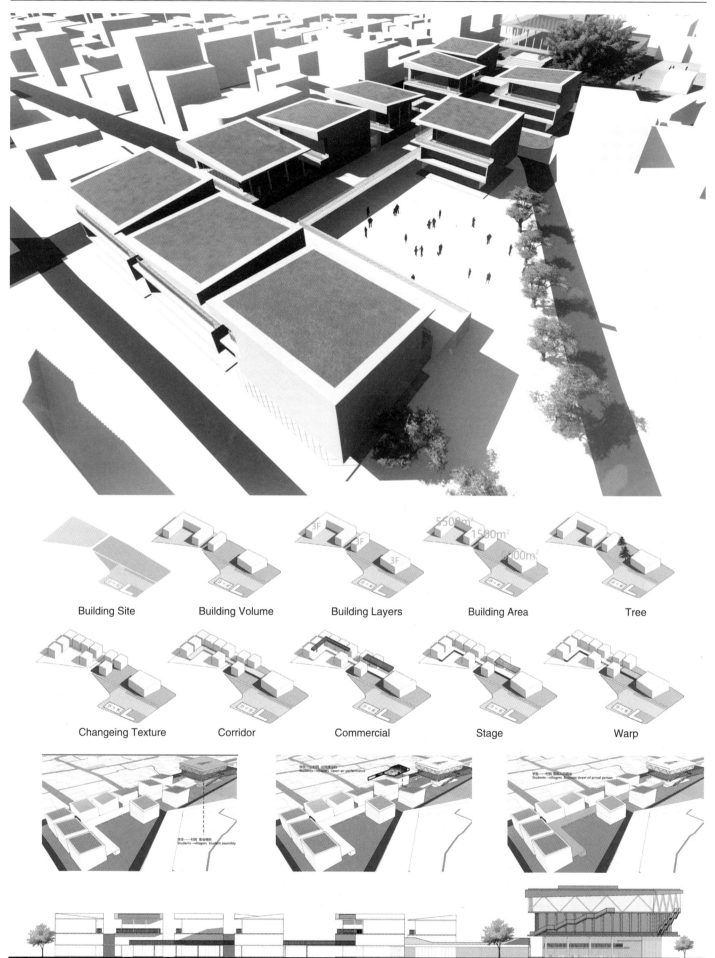

West Facde

融合·共生与介入·整合 —— 华侨大学厦门校区与村落的城市设计研究

2016国际五校研究生联合设计过程回顾
2016 International Joint Studio for Five Universities Review

开幕讲座
Opening Lectures

现场调研
Site Investigation

上篇：融合与共生 —— 华侨大学厦门校区校园空间可持续发展概念设计

小组讨论
Group Discussion

调研汇报
Investigation Report

视频课程
Video Courses

答辩
Reply

相关学术讲座
The Academic Forum

边经卫 教授

讲座题目
《集美新城转型路径对华侨大学厦门校区发展的思考》

许金顶 教授

讲座题目
《侨乡社会历史文化调研及其思考》

陈志宏 教授

讲座题目
《闽南侨乡近代建筑遗产保护研究》

许懋彦 教授

讲座题目
《我国大学校园规划变迁史简述》

古谷诚章 教授

讲座题目
《令人心情愉悦的场所》

费迎庆 副教授

讲座题目
《鹭江社区剧场设计等》

龙元 教授

讲座题目
《城市非正规性视野下的社区营造》

刘塨 教授

讲座题目
《"融合共生"——华侨大学厦门校区校园空间可持续发展概念设计》

刘佳燕 副教授

讲座题目
《从"双营"到双赢：社区营造和城市营销》

成丽 副教授

讲座题目
《华侨大学周边及兑山村历史建筑简介》

2016 国际五校研究生联合设计教学总结会会议记录
2016 International Joint Studio for Five Universities Teaching Plan

一、会议主题："融合与共生"——华侨大学厦门校区校园空间可持续发展概念设计教学总结研讨

二、会议时间：2016年10月15日

三、会议地点：福建厦门京闽北海湾酒店

2016国际五校联合设计总结会于10月15日在厦门京闽北海湾酒店举行。会议由华侨大学副校长刘塨教授主持，参与此次会议的有本次联合教学的指导教师清华大学许懋彦教授，中国科学院大学张路峰教授，重庆大学邓蜀阳教授、阎波教授，以及华侨大学郑志教授、姚敏峰副教授、赖世贤老师，特邀参加此次会议的有华侨大学边经卫教授、许金顶教授，一并参加会议的还有本次参与课程设计的华侨大学全体研究生。会议总结了2016年国际五校联合设计教学成果，并对2017年五校联合设计选题进行了探讨。

以下为会议主要讨论内容（根据会议录音整理）

【刘塨教授】：首先欢迎各位老师来到厦门参加我们这次国际五校联合设计的总结会，今天会议正式开始。我觉得通过这次联合教学我们的收获非常大。学生们很辛苦，不知道他们的收获怎样，但我自己的收获是非常大的。厦门校区的规划从2003年选址开始一直到这次联合设计成果汇报之前，我始终有一个心结——就是那片待拆迁的房子。总感觉把它拆掉一点，华大就会完整一点，而在拆完之前，华大始终是残缺不全的。这次联合设计的题目正是为解决这个问题而设定的。

在听学生汇报成果时，我观察到张路峰老师一边听一边把学生成果的特点以简图的形式记录下来，我觉得这样很好，然后我也跟着画。在画完最后一张图时，心情豁然开朗，之前总是在想只要还有一个没拆迁的房子，华大就是残缺不全的。这次画到最后发现，既然允许我们校区有个小山包保留不动，为什么不能有一个社会型的斑块也保留在那里？我们只要把整个脉络疏通清楚，它就是一个有机的整体，不同的异质型斑块反而会给校园带来新的活力，就像我们校园北部的小山包一样，我们一点都不能碰它，它却成为我们学校的一个特点。因此，对于我们不能碰的东西，只要我们把它的结构疏通好，它就能满足整体有机结构的要求，而且还会成为人居环境的特色。这意味着，我找到存在无法拆迁区域时实现校园规划的理想模式。这是我搞这个规划以来纠结十多年的问题，第一次找到一个学术的答案。在此，我对景观生态学做一个建筑学的引申：在人居环境规划建设中，社会性异质斑块在一定条件下也符合镶嵌性原理。这个原理的拓展，不仅对华侨大学厦门校区的村校融合有意义，而且对处理城中村乃至更普遍的城市更新问题都有参考意义。所以我觉得就这个课题来讲，首先我就很有收获。

【张路峰教授】：首先感谢华侨大学以刘校长为首的团队在整个承办过程中对合作教学的投入与支持。我的总结分为三大部分：1）课题；2）教学本身；3）下一年的选题。

首先说第一个部分——课题。经过了将近一年的教学过程、现场踏勘，包括课堂研讨以及与老师们的一些交流，还有很主要的是课程之外的我们请来的两个老师在不同的角度给我们一些对项目的启发。其实我个人也是跟刘校长一样处于一直在思考的状态，甚至在指导学生的过程中自己也在做这道题。对于这道题到底该怎么做，经过这次合作教学有一些个人的持续思考，也有一些初步认识，我大概总结了六点，可能也不是太严密。

第一点，就是我们所探讨的这个校中村问题实际上就是城中村的问题，城中村问题放到一个城市化的大背景中或者是城市化的过程中去认识才能够找到答案。城中村问题本来在城市研究领域就是一个难解的题，我看了很多这方面的东西，没有一个明确的解决办法。它的一个基本的机制就是城市无序扩张的情况下吞噬了农村的土地，失去土地的农民，实际上就失去了他赖以生存的生活生产资料，他们会夹在城市的"缝隙"中成为弱势群体。他们的生活模式主要是靠出租房屋这么一种我们叫"瓦片经济"的现实。实际上他们过的是一种临时状态，一种临时生活，未来是怎么样的他们不清楚，也没有什么尊严，基本上是城市中最可怜的一帮人。那么这样的一种状态，我们靠规划设计的一种手段、角度能解决问题吗？这显然是比较困难的，因为它涉及很多很多方面，并不是一个单纯的空间问题，空间只是所有这些基质的一种响应。现在城中村问题，我想至少已经有15~20年的时间，之前我就开始关注这件事，但确实没有头绪。现在甚至有的地方就像刘校长说的，心理期望就是要把它干掉，这东西就不应该存在，只不过现在没有实力或者说没有条件把它干掉，但早晚要干掉它，都是这么一个心态。我们面临对城市化这件事本身的一个反思，现在中央政府也已经转变了思想，提出了所谓的"新型城镇化"，新型城镇化的内涵实际上是不再以土地扩张为特征的一种城镇化发展，是一种品质和内涵的建设和提升。在这个大背景下我们如何去修复已经失衡的城乡关系，给城中村一个永久的定义，这可能是非常有意义、有价值的一个选题。我们以学校作为一个例子，校园扩张就好比城市扩张，遭遇拆迁成本造成的村庄夹在校区中的一种现状，其实就形成了一种僵持和对峙的状态，这种僵局怎么打破？对于我来说这件事是非常有学术意义、具有学术价值的，它是一个学术课题而不仅仅是为刘校长解决了一个萦绕十年的工作难题。

第二点，校园和村落这两个看似有根本性矛盾，水火不相容的一种状态，也就是刚才讲的城乡关系的一种反应，这两者有没有可能共融？如果有，那这两者共融的基础在哪？我们经过跟许金顶老师的几次聊天，还有许老师参与我们的教学指导，有了一个认识：实际上这两者是存在共融基础的——那就是华侨文化。华侨大学和华侨村两者的最大公约数就是华侨文化，两者存在共融的基础，那么只要我们弘扬和保护华侨文化，实际上就是在延续我们华侨大学的办学精神，两者的目标是一致的，这也是两者联系的一个纽带。那华侨文化到村落层次它体现在什么层面呢？我认为华侨文化的物质载体就是村落，如果没有村落，我们所讲的华侨文化也就不存在了，在这一点上我们要把保护村落就是保护华侨文化这样一个认识提到一个高度，这样才能够有一种文化上的自觉，才能真正把村落保护上升到跟我们的校园建设一致的目标上。刚才刘校长也讲到，我们在教学的过程中也认识到校园里面有一个村落，我们不应该把它当作一种缺陷或者是缺点，而是应该作为一种特点来思考，从正面去想。这个村落并不是因为它消灭不掉，最后我们不得已地接纳了它，而是因为它带有作为华侨文化载体的一种责任。我们应该非

常庆幸我们的校园中有这样的一个样本，可作为一个村落保护以及华侨文化保护的社会实验室。所以这个村落在校园中的永久性的定位这一点，通过我们课题的探讨，结果是越来越明确了。

第三点，校园和村落的融合共生。这件事若是要操作或者说被接受的话，首先要解决的一个关键的问题就是我们从思想上还是要认识到在校园里保留一个村落，这意味着村落不可能是原生态的村落，校园也不可能是传统意义上的校园。这两者的观念都得发生变化才有可能让二者在一起共存，就像夫妻一样，都得做点改变才可以，否则将两个人硬捏到一块肯定是不行的，也就是说要给对方一个机会，也给自己一个机会，让双方都成为新的事物。规划设计者一定要有创新思维，要改变已有的、固有的一些观念，这件事才有可能实现。哪些方面需要改变呢？首先我们对于高校空间的认识需要改变，我们一定不能再追求那种传统的、封闭式的、衙门式的高校空间。我们对校园空间的传统认识是：一定要有大门，围墙把校园围起来，中间是行政办公大楼，轴线两边是各个学院，就是一种权利为主导的空间形式。我们要借此机会促进校园空间的社会化，这是一种大趋势。我们原来的那种衙门式的空间其实跟我们的教育管理的观念是有关系的。现在这种观念也在逐渐解放，包括好多学院包办的功能也都实现了社会化。把高校作为一种空间资源向社会开放，这件事也是高校未来发展的一个方向，一定要避免再出现那种拆村建校的局面，而是要追求融合共生，这种融合共生不是一种状态的描述，不是最后看上去表面很协调，不是一种视觉上的统一性，这不是我所理解的融合。融合是一定要找到内在机制的融合，它们外在的可能看上去完全不同，但内部发生了一种关联，应该是两者之间产生一种互惠互利互动的关联。所以我认为应该利用保留村落的机会来进行校园整体规划结构的优化和调整，通过保留村落使校园的空间尺度更加人性化，也引入了乡土气息和场所记忆。

第四点，对于村落本身的空间操作层面也有一点认识。我们有一组同学分析了村落建筑的类型，基本上有三种类型的建筑：第一种就是传统民居，包括传统的大厝和洋楼；第二种是村民根据需求自建的小方楼；第三种属于为了赚取更多的拆迁补偿款而建的空壳楼。这三类中第一类应该是保护对象，因为它是稀有资源，带有一定的历史价值，应该尽量以利用为策略；第二种是带有一定实用价值的，就是那种村民自建的小方楼，它的质量也是挺高的。这两种建筑我建议要采取就地保护的策略，不要对它做什么手脚；第三类建筑属于临时性的非法搭建，也没有任何手续，这类建筑本身也跟村民的生活没有什么关系，应该说可以从这些宅基地入手去跟村民谈判，它只涉及利益问题，所以学校可以采取收购、租赁的方式把它的产权界定一下。

第五点，关于整个校园空间结构调整的问题。这一点经过跟刘校长一块画草图也有了一点概念，就是整个校园的东北片区要进行重新规划。原来是从西南角往东北片区蔓延蚕食的状态，"蚕食"到一定程度吃不动就停在了那里，所以东北片区基本上是一个没太搞明白的状态，原有规划也实现不了。新规划要跟西南片区有一定的联系，校园规划本身也是从西南角像年轮一样一层一层往外长的，从规划理念上来讲也可以看作是一种延续性思维。原来是一个岛、一条河、一架桥跨过，就像刘校长讲的所谓自然生态的概念，从水系、人工湿地引领下的空间发展。那么到了这一层的时候，就转变成了拓展的生态的内涵，变成了社会生态的可持续问题。整个校园规划从大的思路上没有发生偏移而是延续深化，从对自然生态的建设转变成对社会生态的建设，它们在理念上是一致的。如果校园中有这么一个新成分存在的话，校园管理模式也要相应地做一些调整，不可能还是按原来那种物业管理模式，可能要更多借鉴市政的方式。

第六点，在操作的策略上建议成立一个校村联合机构来主导校村的融合建设，而不是单方操作。成立一个中间体来协调，要让村民适当参与，至少对这件事有积极性，让他们看到这件事跟他们的切身利益有关，不要让村民整个变成一种被动的状态。因为在被动状态下，你给予的越多，他们越觉得你亏欠他的。首先启动的应该是基础设施和公共服务设施，因为这两个部分的差距是最大的，村里面污水横流臭气熏天，如果能把这种状态改变了，村民们肯定是不会拒绝的。

第二大方面关于教学本身。这次五校联合教学是若干年教学中的一个体验，对于我来说，我每年都在思考一个问题：研究生的设计应该达到怎样一个深度？我们以前做的大多是本科生的教学，现在我感觉对于研究生的设计指导还是没有脱离对本科生毕业设计的指导模式，或者说深度没有明显的进步。我觉得我们还要继续探讨我们做这个联合教学的目标，是要探讨出来研究生的课程设计应该做些什么，应该怎么做，做到什么程度。现在研究生设计与本科生毕业设计深度拉不开距离的原因我认为有三个：1）现场调研太少；2）研究的问题比较笼统、比较大；3）统一的要求限制了各校的特点，我们应该增加一些自主性。第二个教学上的感受就是我们的学生说得多做得少，都特别会讲 PPT，会讲故事，但真正落实到方案上的想法很少。现在的学生大概都是这样，缺乏一个画图的过程，没有在图上思考的过程，我们能否通过本次教学来回归传统的教学理念，像传统的图纸和模型的工作方式，我感觉是更接近建筑的本质的。

第三大方面关于明年的选题方向。我认为有两种可能：一种就是延续今年的课题，今年算是第一轮设计，明年我们继续深化它；另一种可能就是开辟新的课题，甚至是换地方，像北京、重庆也有很多值得我们去探讨的课题。

【许懋彦教授】：第一点关于我们教学本身。我觉得我们这个题目本身，包括一些台湾的老师都认为，新的校园规划模式发展到现在这个阶段以来，我们一直都在谈的是新校区里的规划结构一类的问题，而反过来说在校园里涵盖一个村落的情况，目前国内其他的校园规划中基本上没有讨论过。从这个意义上来说，它是一个新的问题，一个新的阶段，新的探讨，我想这是我们这次联合教学的一个定位。所以大家很关心：我们做这么一个东西，刘校长介意吗？当然不会，因为这是刘校长牵头来做的一件事。那其他校长们能接受我们的设计吗？我想这正是我们要达到的一个预期的目的，不仅仅是影响到校长们，甚至还要影响到区长甚至更高层的领导，使我们的教学结果不仅仅是一个空谈，而是真正的能影响到校园的未来发展。

第二点关于下年的选题。在明年选题这一方面呢，我个人建议如果能深化这个题目的话，我们还是继续做这个题目，如果我们下一届的同学能在现在这些作业思路的基础上，去深入细化某一点，我觉得不仅对学生，对我们这个联合教学都是很有必要的。

【邓蜀阳教授】：我今天的话题跟张老师的有一点不一样，我觉得既然我们是联合教学的一个总结会，那我还是要更侧重于教学的方面来进行思考。

第一点对于选题来讲，选题对教学来说是第一位的，选题的主要目标，实际上都是通过对焦点认知的一种话题式的思考，来进行的一种特征性的选题，也就是说，在这个过程中让大家对这些问题的存在和今后发展的对策以及方向性的思考进行的一种大的研究性的教学工作。我们今年这个校村结合的话题是教学和社会存在融合性的话题，但关于这个教学的焦点，它不一定就只能是校村结合的问题，假如说我们今年选择校村的问题，明年选择废旧工厂的问题，后年选择滨水自然景区的问题，那讨论的方式就不一样了，因此我觉得选题方面实际上就是根据我们社会发展当中所面临的一些问题或者是困难或者是现象，我们可以提出一种思考的话题，这种话题可能是我们今后讨论的一种方向。

第二点就是关于联合教学的组织上的一些问题，因为是联合教学，所以我认为首先就是要突出"联合"两个字，既然联合那要采取怎么样的联合？这是要好好考虑的。我们今年做这个联合教学的过程实际上也正是我们不断摸索完善的一个过程，包括前期的调研工作，调研工作中我们做得很好的一点就是我们把各校的学生进行了混合编组，这样做的一个好处就是可以把各校学生的特点都给发挥出来。在这个过程中，几个学校之间，学生与学生之间，老师与老师之间进行融合，当然老师的融合度往往要比学生之间的融合度高一些，但学生在这个过程中也会得到更多的收获，跟其他学校学生进行一些思想上的碰撞，以及在这个过程中出现的一些交流甚至是碰撞出的火花都会使我们的学生受益。还有我们后期评图讨论时所用的网络平台的模式，这一点也是我们联合教学的一个摸索，我认为还是起到了一个很好的作用的，因为我们各自回到各自的地方后，想再碰个面就很难了，要想再召集一次会议是要付出很多的时间以及金钱上的代价的，那么我们利用这个网络平台为我们实现了一种低成本高效率的交流。但是我总是觉得我们在形式上看起来是融合了，但还是一种比较"粗浅"的融合，层度不够。所以我认为我们应该在第一阶段，在我们调研之前先做一个预研究，让学生带着问题去调研，这样在之后的交流中也能得到一些实质性的结果。

第三点就是关于明年选题，我认为如果我们要继续研究今年的方向，就要做一些细节和片段的深化。

【阎波教授】：简单讲一下我在这次联合教学中的体会。我当时看到这个选题，就觉得这个题很难很大，所以我当时就跟学生说这个题很不容易做，这是我讲的第一点。第二点我想讲一下我们这次联合教学的几个"多"，就是多学科、多校、多方式，这是我们所体现出来的特点，特别是许金顶老师以及清华大学刘佳燕老师在网络平台上给我们进行的教学，使我获益良多。另外我们的汇报方式以及我们的现场调研、网络汇报以及最后的联合评图等这些方式使我们的联合教学不断地完善。在这次联合教学当中我们有这么几个收获：

1）从学生角度来讲，我觉得通过这个教学过程，学生从调研方法，从城市设计方法还有从建筑设计的方法，以及众多学科的比较、多地域的比较，从社会学、文化学等方面，开了很大的眼界。因为这个题目是一个很宏大的题目，所以这也造成了我们学生选题的多层次化，就像许老师在PPT中选取了八个方向来讲评学生作业，其实是远远不止八个方向的，总的来讲我们让学生获得了一个认识：我们传统的文化和精神必须要传承下去，而不能是简单地、粗暴地把它抛弃。

2）对于我自己来说，通过这次联合教学我也有很多的收获。第一点通过其他学科、其他教师不同方式的知识的传授，让我对这个课题有了更加浓厚的兴趣，开拓了自己的视野，对城市设计也增加了一些更深刻的认识，特别是中国科学院大学的几个同学给我留下了很深刻的印象，虽然说他们有些做法我不是很赞同，但总的来说是开拓了我的眼界，让我知道城市设计原来还能这样"玩"，我觉得很特别。另外一个是对传统建筑的调研以及对现场的调研，让我更加深刻地认识了向传统学习的方式，特别是通过中日联合教学这么一个方式。还有一点就是我觉得我们应该向匠人学习，学习匠人的精神和品质。在这个过程中我一直在思考：建筑师的培养以及我们这次联合教学怎么进一步去深入这个话题，结合前面几位教授的讲话，我觉得我们是否可以在选题的过程中向精细化发展。

当然在我们的教学过程中我们也发现了许多问题，比如我们在下一步的联合教学方面是不是可以以不同的目标、不同的方向、不同的联合方式结合起来；比如说现在在我们在调研时实现了多校之间的融合，但现在有一个问题就是学生在调研时属于一个分组，回校后又分到另一个组，这样一来他在前期调研的一些目标、一些想法以及一些预研究就很难展现出来。所以我想我们有没有可能在联合教学的多联合方式上再进一步地深入。

【边经卫教授】：这次联合教学确实很有成效，这个成效一方面是对几个老师所带队的各个学校的研究生的一个锻炼，对我们华侨大学来说呢，如何破解十年来所困扰的问题提出了一些理念和方法；另一方面，我们现在学生的联合设计也好，毕业设计也好，阶段设计也好，感觉我们老师的课题真正结合问题导向的非常少，所以我觉得这种问题导向式的联合设计非常好。当时一看到这个题目我感觉很兴奋，我想不论学生做得怎么样，首先这个思路给我的感觉就是一个很好的思路，这种真正的体现问题导向式的联合设计让学生真正地融入社会实践、社会活动，来提升学生真刀真枪做事的能力，在这方面我相信这批学生一定是有一个深刻的感受的。

在整个联合教学的过程中我有这么两点理念：第一个理念就是我在想我们接下来做一些事情能不能按照城市修复的理念来做，所谓城市修复就是要保护原有的老的基质，但是保护不代表什么都不能动，怎么样来实现新老的巧妙地结合，是我们应该去思考的东西。兑山村实际上就是一个城市修复的好例子，如何把兑山村和我们的校园进行一个巧妙的连接，使它达到一个融合共生的目标，我想是可以利用城市修复、乡村修复这么一个概念来做。

第二个理念就是城市设计的理念，刚才几位老师也都有谈到我们今后怎么来做这个东西。首先继续做这个课题我认为是非常有意义的一件事，如果要继续往下做，我还是建议从城市修复的角度、从城市设计的角度来深入，不仅仅是只针对校中村的修复，还可以包括我们现有校园已建的一些不合理的地方来进行修复式设计，真正地使我们的校园结合兑山村形成一个大校园的概念，而不是仅仅停留在表面的一种融合。

【许金顶教授】：在以前历史学跟建筑学应该说是死对头，因为说实在的，后现代主义就是从哲学跟建筑学开始的，建筑学追求的是一种不要有传统的评论，不要传统的价值，而是个性的、多元的，每个人有每个人的理解，所以这样的后现代主义对于我们原来见过的文化传统伤害很大。西方近几年对于后现代的反思虽然也挽回了一点对传统的伤害，但都是很微弱的。当时刘校长让我来参加这个五校联合设计，我十分惶恐，我在想我该怎么去参与到这个教学过程中，但当我看到这个题目时我就明白了这就是我们共同追求的一种状态，刚好这些村庄我也有些资料，就拿出来跟大家分享一下。我从这个联合教学的过程中也学到了很多东西，我没想到学生能设计出那些超乎我想象的作品。那么就这一次参与这个活动本身，我有几点认识：我非常赞成张路峰老师讲的这个校村要联合必须要有一个协调机制，如果没有这个协调机制，很容易形成一种矛盾的深化与对抗，关键就在于这个协调机制建立的价值观和理念是什么，这就涉及非常非常多的利益平衡的事情。村民有一个底线，就是村里的宗祠和神庙你不能动，他们是要回来拜神祭祖的，这就可以很好地回到文化原点这个概念上。家庙、神庙、宗祠、洋楼你不要去动它，古榕树也不能动，在这一点上我们华侨大学现阶段做得非常不错，可以看到原来村里的大榕树我们都给它保留了下来，而且保护得非常不错。在这次联合设计中，我们的学生也表现得不错，把这些文化原点的东西也都保留了下来。那问题也就随之而来了，如果我们把这些东西保留了下来，而传统的活动却都消失了，那这些也就只是破房子一个了，所以我非常赞同张路峰老师讲的要有人的存在的这么一个理念。村民虽然住出去了，但我们要保留他们回来拜神祭祖的权利，也就是校园的开放性，村民回来举行活动实际上就是侨文化的一种展现和延续，我们学校应该积极地配合他们，甚至我们的学生可以去给他们当志愿者，从而参与到活动中。

关于明年选题，我不希望这个课题今年刚开始明年就停了，我倒是觉得可以做一个2、3年的一种反复的实践过程，每年的角度可以换一下，我们把这个课题给做圆了，做实在了，这样再来总结教学经验和未来的选题，我们才有更多的内涵来发挥，如果今年我们就换课题，那就变成了在很不扎实的基础上去做比较。

【刘塨教授】：这个项目我在刚才开始的时候提到过，我作为一个学校管理者的角度，有很大的收获，解决了很大的难题。这个难题首先是在学术上的解决，我觉得这一点很好，因为连学术上都解决不了的事情，却去谈政策上如何解决，这样我在心理上从最根本上是不认同的。但现在在学术上这个问题打通了，剩下的就是一个策略方法的问题，即怎么逐步地实施与落实，所以这个是我最大的收获与感受。毕竟这个心结存在十多年了，真的是很难受，总觉得自己一个完美的设计卡在那里就是搞不定，那现在这个问题基本上是解决了。

回过头来讲在教学过程中总结到的几个面：第一点，我同意大家提到的我们这个题目继续往下做，这是我基于对这个题目的设置以及我们研究生的设计课程设题的问题的认知这样说的，因为我想我们本科生的毕业设计也讲研究，也要调研，那它跟我们研究生的设计差别在哪里呢？从这个题目做的过程中体现的差别我觉得还是很大的。我个人的理解，本科生的调研、设计，它研究的问题是个案问题，严格地说它研究的问题在学术上已经没有问题了，老师对他的指导和给他课题的设置实际上是在学术上有明确的结论的、定式的东西，是个案问题，可能地形、设计要求有所不同，本科生要调研的是这么一种情况；而研究生的研究设计，实际上首先它在学术上必须是一个有待研究的问题，还没有一个套路的、成熟的解决方案，要求研究生首先在学术上创新，然后再放到方案上创新，而且，这两个创新必须是连续的。而我们这次的题目设计正是这样，首先我们讲这个城中村、校中村它不是个案问题，而是一个很普遍的学术问题，而且在国内特别是未来的城市发展到新阶段的情况下，这个学术问题会越来越突出——它现在已经很突出了，现在已经涉及国家的发展政策方针的问题了。前一阶段的城市发展我觉得可以把它简称为"建设的城市"，就是一个大拆大建的建设的过程，是一个"建筑学的城市"。未来我感觉可以把它概括为是一个"回归社会变迁的城市"，是一个"社会学的城市"，而不再是那种大拆大建，恨不得几天就搞定一个新区的城市化，是在社会的正常节奏变迁下的城市化，所以我们这个课题在学术上有非常大的容量，以至于我们这个题目看起来偏大了。其实我们开始想做的事是几个学生一组做城市设计，然后每个人再做单体建筑设计。实际上我们最后的结论是，其实大家都没有做到这么一种深度，所谓的单体设计还是在城市设计的深度上面，根本做不到下一个阶段，为什么？因为前面一个阶段已经把大家搞得筋疲力尽了，首先这个学术问题对于我们老师来说就需要探讨，所以我们在教学的引导上不可能给予充分、系统的指导，因为它对于我们来说也是一个课题，我们也要研究。所以正是从这一点我们至少也要搞2、3次，我们自己把这个学术问题先搞清楚，这样的话才能真正达到一个教学的效果，不论学生做得怎么样，至少我们对我们要教的东西搞清楚了，把这个大问题给它搞明白，所以我觉得我们应该继续深入做下去，这个学术问题有绝对大的容量，不仅仅是对我们研究生来说，对我们老师的学术研究也是这样。

第二点，我是觉得面对这样的问题，城市设计的一种理论介入不仅是可能的，而且是必须的。这样的城市问题不可能在建筑上得到解决，我们一定要从狭隘的建筑学走向兼容社会学的一种大的建筑学，就像刚才大家提到的大校园的概念，我也可叫它大侨村，村民也不要排斥学校，我们学校是一个侨校，跟大家都是在一个大的文化平台上的，所以把社会学兼容之后我们才能真正地搞城市设计。建筑设计可以是建设的过程，但城市设计它一定是跟这个城市变迁相结合的，不能通过拆了建的方式来形成一个东西，实际上倒是要用这种聚落的发展、社区的营造这样的一种概念去搞我们的设计。通过这样的过程，我们反过来也更能认识到城市设计的特点，也是从纯粹的空间的营造走上社会发展的节奏的东西。

那么对我们这个课题，如果继续往下做的话，我认为我们可以有两种方式。一种是学术路径方向，比方说做水的，比方说强调侨文化的；还有一种我觉得可以是通过几个方面，比方说我们现在校跟村目前还是一种比较粗暴的对接方式，为了学校的安全稳定，管理方面就用一堵墙把它隔开，那学校的环境和村里的环境这个界面怎么融合，就是一个具体的项目了。那么村落保护的特点和校园教育环境的特点怎么来结合？是一个界面还是一个地带？界面该怎么做？地带又该怎么做？这些具体的问题也就出来了，这样我们就会有更专门性的调研指向，从而进行一些更深入的空间运作，以及能够符合这种社会学变迁的思路。这样的话，我们对这个课题本身在学术的认识上就会更加系统地深入下去，对我们整个的教学也是一个完善。

2017 国际五校研究生联合设计课程
2017 INTERNATIONAL JOINT STUDIO FOR FIVE UNIVERSITIES

下篇
PART TWO

介入与整合

基于校村共生的华侨文化保护与侨村环境整治方案设计

INTERVENTION AND INTEGRATION
PRESERVATION AND REVITALIZATION DESIGN FOR THE
TRADITIONAL VILLAGE-WITHIN-THE CAMPUS IN HUAQIAO UNIVERSITY

2017 国际五校研究生联合设计教学计划
2017 International Joint Studio for Five Universities Teaching Plan

一、课题背景

中国在快速城市化的进程中，空间扩张速度远远超出社会发展节奏，政府主导的基础设施网络急剧扩张切割原有村落肌体，土地征收与房地产开发使得大量传统村落迅速消失、衰落。土地投机的随意性与拆迁补偿的复杂性，造成"城中村"等城乡交错的普遍现象。

与此同时，城市利用高校大扩招，纷纷建设大学城、文教区。是我国高等教育由精英教育发展进入大众教育的重要阶段，给各高校带来巨大的发展资源。

华侨大学厦门校区建设使学校发展空间翻倍，在学生规模扩大60%的前提下，彻底解决严重超标的拥挤不堪。同时师资引进、生源提高等方面获得巨大动力。大学城、文教区加速了城市化，但遗患匪浅。宏观上巨大的斑块尺度和相对单一的功能，造成严重的生态失衡，将长时间游离于城市的生态律动与整合。微观上又造成村校混杂交错的"村中校、校中村"的复杂状况。二者空间脉络上相互排斥，但经济上又相互依存：村中的"学生街"为学生提供日常生活服务，学校的物业、保安、保清岗位为村民就业提供了机会，形成了村民与教师共跳广场舞的局面。如何将二者的关系从功能到形态进行整合，建立起和谐的融合与共生关系，是一个巨大的挑战。

二、校园概况

华侨大学1960年创办，直属国侨办。是一所面向海外、面向港澳台，为侨服务，为经济建设与社会发展服务，传播中华文化的国家重点建设的综合性大学。现有全日制在校生28000余人，平均分布在泉州、厦门两个校区。

华侨大学厦门校区位于集美文教区，集美大道北侧，背靠天马山面对杏林湖。规划总用地2000亩，总建筑面积63万平方米，"十一五"规划学生数15000人，远期20000人。2004年11月开工建设，2006年10月投入使用，现完成规划面积1300亩，建筑面积55万平方米，全日制在校生14000人。

校园规划突出生态理念，一是主体行为生态，把学生生活区布置在中间，到各点的距离在400米左右，故全校园以步行为主，骑自行车者不足3%。二是环境行为生态，突出体现在水环境，打通校园与周边的生态脉络，形成基于"功能湿地"的生态校园，不仅实现海绵城市功能，还实现污水零排放、节水1500吨/日、水禽保护地等众多生态效益。

建设基地原属兑山村，分为潘涂、下蔡、宅内、祖厝后、西头、西珩、南尾井七个自然村落。根据国侨办和厦门市政府的战略合作协议，拆迁由当地政府承担。拆迁现状，红线内剩下场地东南的潘涂、宅内、下蔡，约600农户，西北的南尾井、西珩，约100农户。校园中间还有一户"钉子户"，是典型的村校错落状态。

根据校园规划，华侨大学厦门校区分为教学区、生活区、文博区与运动区四个分区。教学区、生活区基本建设完成，运动区部分完成，"文博区"尚未开始建设。本课题所研究的范围即是原校园规划校园总用地范围(2000亩红线)，考虑课题研究对象村庄的完整性，可适当扩大至校园规划外的部分。

三、课题释义

本次课题的核心议题是探讨校村共生的前提下，如何通过有限介入，激发侨乡人居环境的有机更新与发展，使生存环境得到提升与改善，华侨文化得到保护与延续。

兑山村华侨文化源远流长，特色浓郁。对于华侨大学的办学宗旨来说是极为独特的宝贵资源。然而侨乡传统的生存方式跟不上现代的节奏，整体趋于衰落。而校园的出现与急剧拓展瓦解了原有的生产方式与空间网络，在空间肌理上形成极大的对峙与隔阂，解体消融还是更新发展？侨乡走到了十字路口。

另一方面，在空间对峙与隔阂中，社会意义的融合与共生也在悄悄发生：学校物业为村民提供了新的广泛就业机会，校园的很多空间成为村民休闲与活动的场所，村里的商业与公共服务也成了师生的多样化选择并为之注入新的活力，让我们看到侨乡有机更新的希望与潜能。

侨乡与侨校存在文化深层的同一性，通过侨乡的有机更新，目前两者空间的对峙与隔阂一定会消除，最终使侨村侨校达成完美的融合。我们要做的是推动侨乡自我生长、自我更新，让侨乡按照自己的节奏与规律成长，这样才能使华侨文化作为活体得到保存与发展。这样的介入，方式如何？要素如何？过程如何？都是需要我们认真深入思考和研究的重点。

拟有以下研究方向：
1) 适应校村共融发展的村落有机更新的结构介入；
2) 适合社会需求的村落功能的调整与改造介入；
3) 适应"侨村"与"学村"特殊定位的调整与改造介入；
4) 侨校与侨村中特色文化景观的强化介入。

有以下几个建筑设计方向可供选择：大学生活动中心、博物馆、华侨大学校史馆、音乐厅、休闲咖啡厅、综合服务楼。

四、教学目标

引导学生思考在新型城镇化的大背景下传统人居环境有机更新模式；

引导学生理解"生态校园"的可持续规划理念，理解"自然生态"与"社会生态"可持续发展的意义；

训练学生从城市尺度理解校园与村落空间，学习以微观的空间操作手法介入和干预一个复杂社会空间的方法和技能；

要求学生掌握文献调研与现场调研的方法。

五、教学阶段与任务描述

本学期教学分四阶段进行。各阶段任务与成果要求如下：

第一阶段 本阶段为"预研究"（Pre-Research）阶段，（《预研究专题举例》附后，供各校参考）要求各校内小组合作进行现场研究回顾与现状分析，文献研究与案例分析，理论阅读，提出总体设计策略与目标。

第二阶段 本阶段为"现场研究"（Site Survey）阶段。该阶段要求集中在现场进行实地调研和踏勘，五校师生混合编组，分不同专项进行调研工作（调研提纲小组自行拟定）。

第三阶段 本阶段为"概念设计"（Conceptual Design）阶段。

各校内小组合作针对整个研究地块提出总体性、概念性规划设计方案。

第四阶段 本阶段为"深化设计"（Design Development）阶段。小组深化总体设计方案；个人选择与总体设计意图相关的单体、群体、空间节点或系统进行个体介入设计，完成设计成果并进行联合答辩。

第五阶段 各校分别整理完善成果，准备出版展览文件，校内答辩展示交流。

1. Background

In the process of rapid urbanization in China, the space expansion speed is far beyond the pace of social development, and the original village body is cut by the sharp infrastructure network expansion led by the government, and land expropriation and real estate development has made a large number of traditional villages rapidly disappear and decline. In addition, randomness of land speculation and complexity of demolition compensation has caused the general urban-rural crisscross phenomenon, like village in city.

Meanwhile, the city constructs the university city as well as culture and education area using the large enrollment expansion of universities, which is an important stage of higher education from elite education to mass education, bringing huge development resources each university.

Construction of Xiamen campus of Huaqiao University doubles the school development space, with 60% scale of student expansion, solving serious crowding and providing great motivation of teacher introduction and enrollment enhancement. The university city as well as culture and education area have accelerated the urbanization, with deep problems left. Macroscopically, the huge plaque scale and single function cause serious ecological imbalance with ecological rhythm and integration for a long time. Microscopically, it causes the staggered village – within – campus and campus – within –village. There is space conflict and economy interdependence between them: the student street in the village can provide daily life services for students, and property, security and cleaning in the university can provide the employment opportunity for villagers, forming a situation of joint square dance of villagers and teachers. It is a huge challenge to integrate the relationship between them from function to form, to build a harmonious relationship of integration and symbiosis.

2. School Profile

Huaqiao University, founded in 1960, directly under the Overseas Chinese Affairs Office of the State Council, facing overseas, Hong Kong, Macao and Taiwan, servicing for overseas Chinese and economic construction and social development, and spreading Chinese culture, is a key comprehensive university. There are 28000 full-time students in Quanzhou campus and Xiamen campus on average.

Xiamen campus, located in Jimei Culture and Education District and north of Jimei Avenue, backs against Tianma Mountain and faces Xinglin Lake. The total planned land is 2000 mu, total construction area is 63 square meters, number of students of the eleventh five-year plan is 15000 and 20000 at a specified future date. Construction began in November 2004, which was put into use in October 2006, and now the completed planning area is 1300 mu, construction area is 55 square meters, and the number of full-time students is 14000.

The campus planning can highlight the ecological idea: first, the main body behavior ecology: the student living area is set in the middle away from each point for 400 meters, so people give priority to walk, and cyclists are less than 3%; second, environment behavior ecology: in water environment, the campus and the surrounding have a connecting ecological context, so the ecological campus is formed based on the function wetland, not only realizing the sponge urban functions, but also realizing wastewater zero discharge, water saving of 1500 ton/ day, waterfowl protection area, and many other ecological benefits.

Construction base originally belongs to the Duishan Village, including Pantu, Xiacai, Zhainei, Zucuotou, Xitou, Xiyan and Nanweijing. According to the strategic cooperation agreement of the Overseas Chinese Affairs Office of the State Council and the Xiamen Municipal Government, the demolition shall be borne by the local government. For the status quo of demolition, there are southeast Pantu, Xiacai and Zhainei with about 600 farmers, northwestern Xiyan and Nanweijing with about 100 farmers, and middle Zucuotou within the red line, as a typical village – within – the campus.

According to the campus planning, Xiamen campus is divided into school area, living area, museum culture area and sports area. The museum culture area has not been completed yet. This topic researches the total land area of planning campus and original campus (2000 mu of red line). Consider the integrity of the research object (village), the research range can be appropriately expand to the part outside the planning campus.

3. Topic Definition

The core topic is to investigate that how to arouse the organic renewal and development of the living environment of the hometown overseas Chinese using limited intervention under the premise of village – within – the campus.

The Duishan Village has a long history with rich characteristics, as very unique and precious resources for the principle of Huaqiao University. However, the way of life can't keep up with the pace of modern, and tends to decline as a whole. The emergence and expansion of the campus sharply disorganizes the original production mode and space network, forming a great deal of confrontation and estrangement on the mechanism of space, so the hometown of overseas Chinese comes into the crossroad of collapse ablation or update development.

In space confrontation and estrangement, on the other hand, the integration and symbiosis with the social significance quietly occurs: school property provides villagers with new jobs widely with a big space for leisure and activities. The business and public service in the village provide teachers and students the diversified chooses, which is injected new vitality for the hope and potential of the organic update.

The deep culture identity of overseas Chinese in hometown and school will achieve their perfect fusion, through the organic update of hometown of overseas Chinese to eliminate confrontation and estrangement. We need to promote its self growth and self-renewal in accordance with its own growth rhythm and regularity to preserve and develop the overseas Chinese culture. And we should consider the intervention way, element and process.

The following research direction:

1）The village structure intervention of the organic update to suit the co-fusion development of school and village;

2）The adjustment and reform intervention of the village function suitable for the social demand;

3）The adjustment and reform intervention of special position suitable for the overseas Chinese village and the school village;

4）The intensive intervention of characteristic culture landscape of the overseas Chinese village and the school village.

4.The Teaching Goal

To guide students to think the organic update mode of the traditional residential environment under the background of new urbanization;

To guide students to understand the concept of sustainable planning of ecological campus, and understand the sustainable development meaning of nature ecology and social ecology;

To train the students to understand the campus and the village space from the city scale, and learn involvement and intervention of a complex social space by microcosmic space operation skill;

To require students to master the methods of documentary research and field investigation .

5. Teaching Stage and Task Description

Stage 1 This is "Pre-Research" stage. Teams in each school should cooperate in conducting site survey review and current status analysis, document research and case study, theory reading and putting forward the overall design strategy and objectives;

Stage 2 This is "Site Survey" stage. Students should do field investigation and exploration on the site. Teams should be comprised by 5 students from different schools, and do research works on different monographic subjects;

Stage 3 This is "Conceptual Design" stage. To put forward the whole conceptual planning and design project by the group aiming at the whole research plot.

Stage 4 This is "design development" stage. The group deepens the overall design plan; the unit, group, space node or system related to the overall design intent and personal choices conduct the individual intervention design, complete the design results and conduct joint reply;

Stage 5 Teams in each school should optimize the design results, prepare documents for publishing exhibitions, and then present the exchanges on campus.

清华大学
TSINGHUA UNIVERSITY

目录 CATALOG

华大八景　唐诗童　丁惟迟
Huaqiao Eight
街环引侨　刘倩君　Germana Isacco
LOOPERS—A Walk to Connect and Develop
生态博物馆　Chiara Menna　马宏涛
The Eco-Museums
合·时序　韦拉　徐淼
Co Order

下篇：介入与整合 —— 基于校村共生的华侨文化保护与侨村环境整治方案设计

学生团队 Student Team

 徐淼

 韦拉

 丁惟迟

 刘倩君

 马宏涛

 唐诗童

 Chiara

 Germana

融合·共生与介入·整合 —— 华侨大学厦门校区与村落的城市设计研究

合·时序
CO ORDER

指导老师：许懋彦
设计者：韦拉 徐淼

区位分析 / Location Analysis

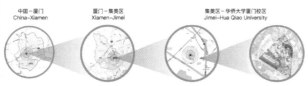

需求分析 / Demand Analysis

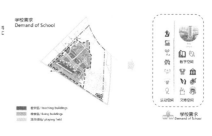

发现问题与概念由来 / Discover Problems and Initial Concept

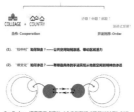

Co-Order, 该方案重点表达"合作与渐进式的开发时序"对于"校中村如何融合"、"侨文化如何渗透"这两个核心问题的理解。想打破村落封闭的情况，外围空间作为村外最容易被察知的空间，最模糊的出发点，通过合作策略的实施，打破校村边界、高边与村落的边界，实现融合。

Co-Order, the project expresses the understanding of the two core issues of "how to integrate the school village" and "how to infiltrate the overseas Chinese culture" by focusing on the "co-operation and progressive development sequence". Inorder to break the situation of closed village, the outer space as the most easy to be perceived space outside the village, is the starting point of the concept. Through the implementation of cooperation strategy, break the boundaries of the school and village, break the boundaries of surrounded area and the village.

基地梳理 / Site Sorting

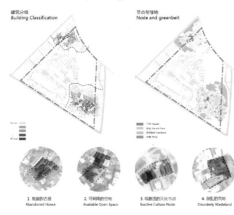

通过对基地内部的建筑分级、节点、绿地以及建筑现状等进行评估，通过类型学的分析，可以看出可以通过利用基地内的废弃的古屋、可利用的空地、待激活的文化节点以及杂乱的荒地等空间进行因地制宜的校中村改造与更新。

Evaluation of the base through the interior of the building, green building, classification, node status, through the analysis of typology, we can see that the village renovation and renewal through space culture within the base nodes using the abandoned bays, available space, to be activated and litter in the wasteland local school.

功能分析 / Function Analysis

教师评语：

校村融合，往往忽视村落而更偏重校园的规划建设。重拾村落振兴成为校村融合的重要课题。华侨大学校园内的村落现状承载着校园的边缘功能，无论从需求还是群体上都存在一定的错位。作品提出"合作与渐进式的开发时序"，以从点到线再到面的方式重新激活村庄的活力，增强与校园的融合。其中，除了对实体设施的更新改造外，强调渐进式的推进与人的活动推广拓展的结合，也是一种有益的尝试。

下篇：介入与整合——基于校村共生的华侨文化保护与侨村环境整治方案设计

节点功能分布图 / Node Function Distribution

实际操作 / Actual Operation

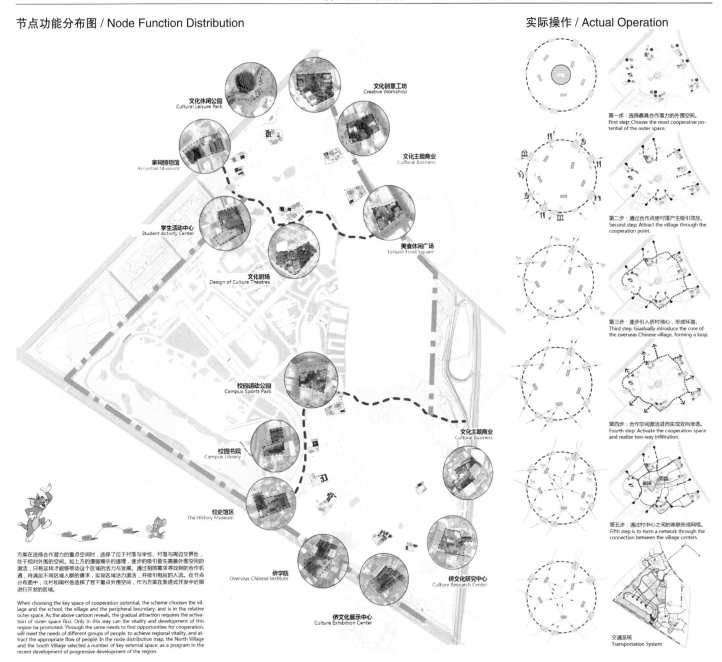

方案在选择合作潜力的重点空间时，选择了位于村落与学校、村落与周边交界处，处于相对外围的空间。如上方的漫画揭示的道理，逐步的吸引首先需要外围空间的激活，只有这样才能够带动这个区域的活力与发展。通过相同需求寻找到新的合作机遇，将满足不同区域人群的需求，实现区域活力激活，并吸引相应的人流。在节点分布图中，北村和南村各选择了若干重点外围空间，作为方案在渐进式开发中近期进行开发的区域。

When choosing the key space of cooperation potential, the scheme chooses the village and the school, the village and the peripheral boundary, and is in the relative outer space. As the above cartoon reveals, the gradual attraction requires the activation of outer space first. Only in this way can the vitality and development of this region be promoted. Through the same needs to find opportunities for cooperation, will meet the needs of different groups of people, to achieve regional vitality, and attract the appropriate flow of people. In the node distribution map, the North Village and the South Village selected a number of key external space, as a program in the recent development of progressive development of the region.

137

下篇：介入与整合 —— 基于校村共生的华侨文化保护与侨村环境整治方案设计

节点设计 / Node Design

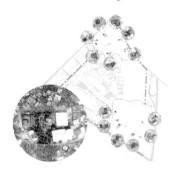

文化创意工坊
Creative Workshop

场地现状散落着荒废的古厝、厂房及大片绿地。设计保留古厝肌理，改造厂房形成学生文创基地，并将绿地公共化，形成新的空间结构。
The site is littered with dilapidated ancient buildings, factories and large tracts of green land. The design preserves the ancient brick skin texture, transforms the workshop, forms the new spatial structure.

文化创意工坊 / Creative Workshop

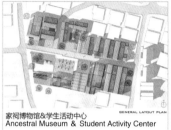

家祠博物馆&学生活动中心
Ancestral Museum & Student Activity Center

场地现状为村落入口空间，宗祠、废弃的古厝以及水池都为改造提供条件，通过肌理梳理、改造，形成入口轴线空间，引入人流。
The site situation for the village entrance space, hall and old houses to provide the conditions for the transformation, form the village entrance axis space, attract the stream of people.

家祠博物馆&学生活动中心 / Ancestral Museum & Student Activity Center

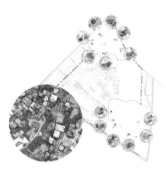

侨学院
Overseas Chinese Culture School

场地现状有家祠的宗祠、废弃老屋的墙面以及方块楼，改造墙面形成半室外展场，保留方块楼结构为成为教学空间，延伸宗祠轴线。
The site situation has family ancestral hall, abandoned walls and block building, metope form semi outdoor exhibition area, retaining structure become the teaching space, extending the axis of the ancestral hall.

侨学院 / Overseas Chinese Culture School

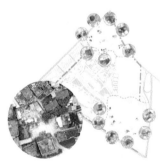

侨文化研究中心
Overseas Chinese Culture Research Center

场地现状为废弃的古厝及废弃的荒地。对现状进行评估，保留大厝，对基地进行重组，形成内聚的中心空间，对外进行渗透，将节点激活。
The site is the abandoned ancient sites and abandoned wasteland. Assess the status quo, retain the Great Wall, reorganize the base, form a cohesive center space, infiltrate the outside and activate the node.

侨文化研究中心 / Overseas Chinese Culture Research Center

重点区域城市设计总平面图 / Core Area Urban Design Master Plan

总平面图 / Mater Plan

融合·共生与介入·整合 —— 华侨大学厦门校区与村落的城市设计研究

校史馆区 & 侨学院 / The School History Museum & Overseas Chinese Culture School

设计说明 / Design Description

方案在靠近校门的古居遗迹区域，通过功能植入、改建、新建等方法，逐步实现将不同的院落进行串联、路径串联，形成用于校史展览、校区介绍、通过实践与参观，进一步建立实践与教学相关的侨学院，成为学术实践、知识获取、教学及文化休闲的核心区域。

In ruins area near the gate, through the function of implantation and reconstruction, new methods, gradually realize the different compound path series, series, used to introduce the history exhibition, and then through the campus, and visit the practice in the old houses in the further establishment of overseas Chinese Institute of practice and teaching related.

场地位置及现状 / Site Location and Status

开发时序 / Development Order

总平面图 / Mater Plan

南立面图 / South Elevation

西立面图 / West Elevation

下篇：介入与整合 —— 基于校村共生的华侨文化保护与侨村环境整治方案设计

节点平面 / Node Plan

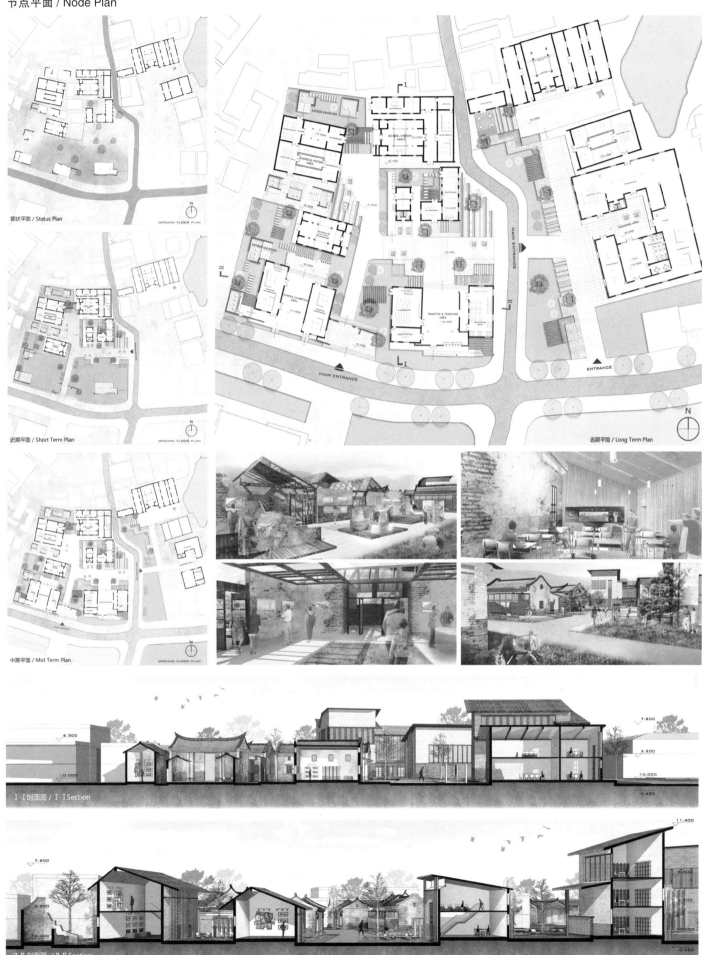

侨文化展示中心 / The Overseas Chinese Culture Exhibition Center

设计说明 / Design Description

方案结合场地的特征，寻觅院落肌理的痕迹，通过渐进式的设计手法，以院落为建筑语言，在保留场地记忆的同时补院落结构，也在时间的发展下植入并完善功能需求，也使得大尺度的肌理向村落进行渗透。激活后的场地对学生市民而言，是体验感知侨文化的窗口，同时也是侨村侨胞文化展示的平台，是其精神内核的墓碑。

Combined with the characteristics of the site, looking for traces of the courtyard texture, through the gradual design style, in the courtyard as the architectural language, while retaining the memory while the dam site courtyard structure, and improve the function of demand also implanted at the time of the development, but also makes large scale texture penetrate into the village.

场地位置及现状 / Site Location and Status

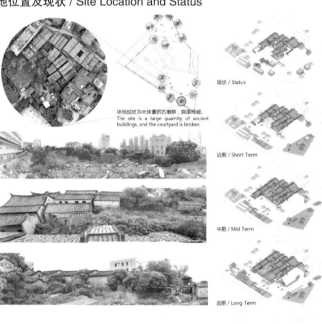

场地现状为大体量的古厝群，院落残破。
The site is a large quantity of ancient buildings, and the courtyard is broken.

现状 / Status
近期 / Short Term
中期 / Mid Term
远期 / Long Term

开发时序 / Development Order

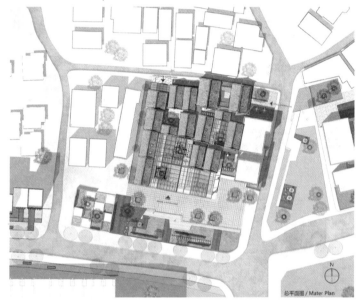

总平面图 / Mater Plan

南立面图 / South Elevation

北立面图 / North Elevation

下篇：介入与整合——基于校村共生的华侨文化保护与侨村环境整治方案设计

节点平面 / Node Plan

融合·共生与介入·整合 —— 华侨大学厦门校区与村落的城市设计研究

生态博物馆
THE ECO-MUSEUMS

指导老师：许懋彦
设计者：Chiara Menna 马宏涛

❶ 书店
❷ 咖啡厅
❸ 社区中心
❹ 传统手工艺展览
❺ 网吧
❻ KTV
❼ 体育馆
❽ 图书馆
❾ 学生街
❿ 纪念馆 & 校史馆
⓫ 剧院
⓬ 艺术中心

华侨大学总图

1	2	3	4
可以不必建筑优先，关键是能否负担起博物馆的基本功能，同时发挥公共性精神文化空间的作用	可以新建，也可以对传统公共空间进行功能置换（例如安娜考斯提亚社区博物馆利用废弃电影院改建）	独特的藏品征集策略和使用方案。藏品应当可通过众筹等方式将社区环境、生活、历史丰富地展示出来	业务活动强调服务社区居民，展览、社会教育、社会服务首先面向居民

教师评语：

"生态博物馆"不是一个具体的实体建筑，它在校园和村落里是一种类似"名片"的设置。作为一个强有力的介入因子，它不仅是对现有资源的保留与陈设，更是对校园以及村落社会网络的提示和引导。通过分散型的规划，在旧有的肌理中形成一张符合现在与未来的校村融合网络。在不同的功能定位中，描绘新的校园生态，激发新的活力与联系。

校园扩建
1 活动室 & 饭店
2 图书馆
3 体育馆
4 实验楼
5 运动场
6 宿舍楼

城市边界
1 纪念馆 & 校史馆
2 剧院
3 艺术中心

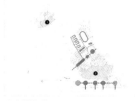

生态博物馆
1 娱乐功能：书店、咖啡店、网吧、KTV、展览、社区中心
2 功能分散的生态博物馆功能

连接
1 入口 & 服务中心
2 古榕公园
3 美食广场
4 游神广场
5 研究机构 & 花园 & 农场
6 室外咖啡 & 展览

道路系统

绿化系统

水系统

传统建筑更新

下篇：介入与整合——基于校村共生的华侨文化保护与侨村环境整治方案设计

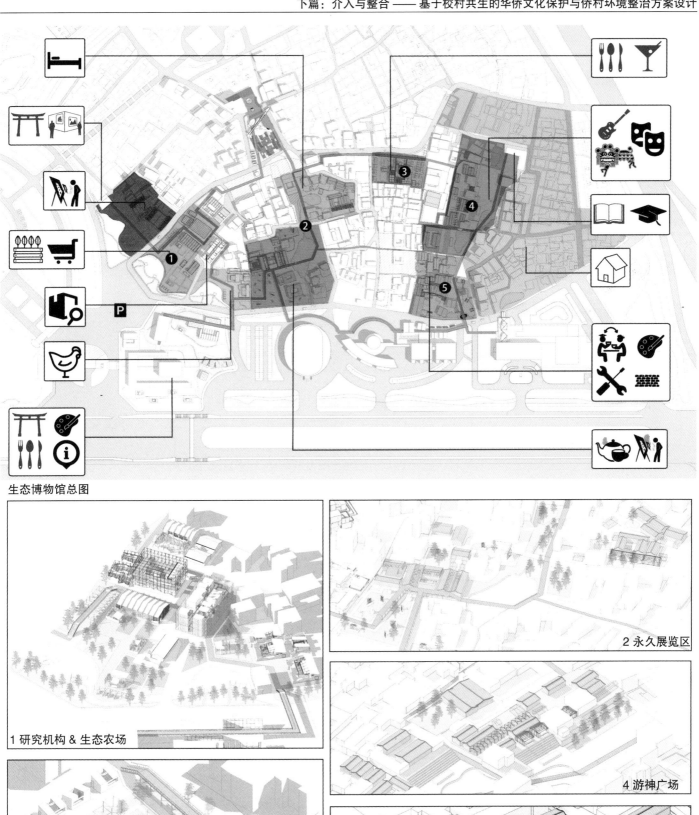

生态博物馆总图

1 研究机构 & 生态农场

2 永久展览区

3 美食广场

4 游神广场

5 艺术工坊

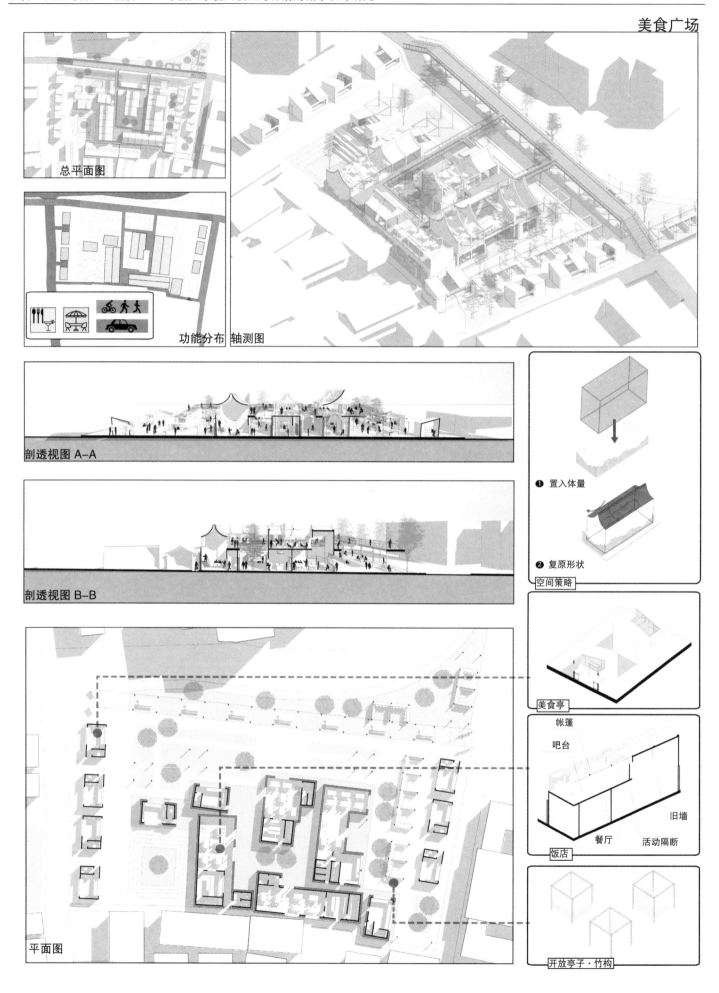

下篇：介入与整合 —— 基于校村共生的华侨文化保护与侨村环境整治方案设计

生态农场

总平面图

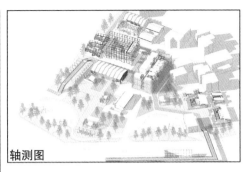
轴测图

功能分布

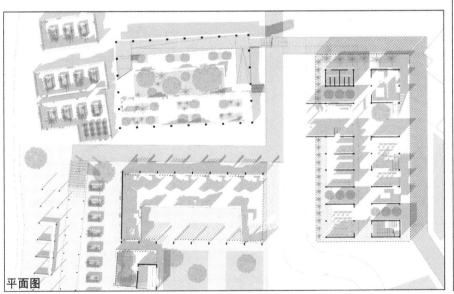
平面图

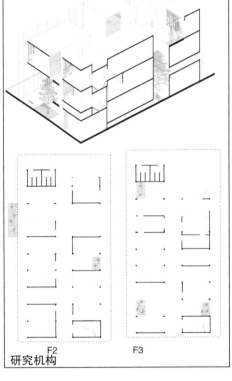
F2　F3
研究机构

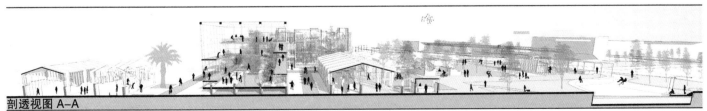
剖透视图 A-A

147

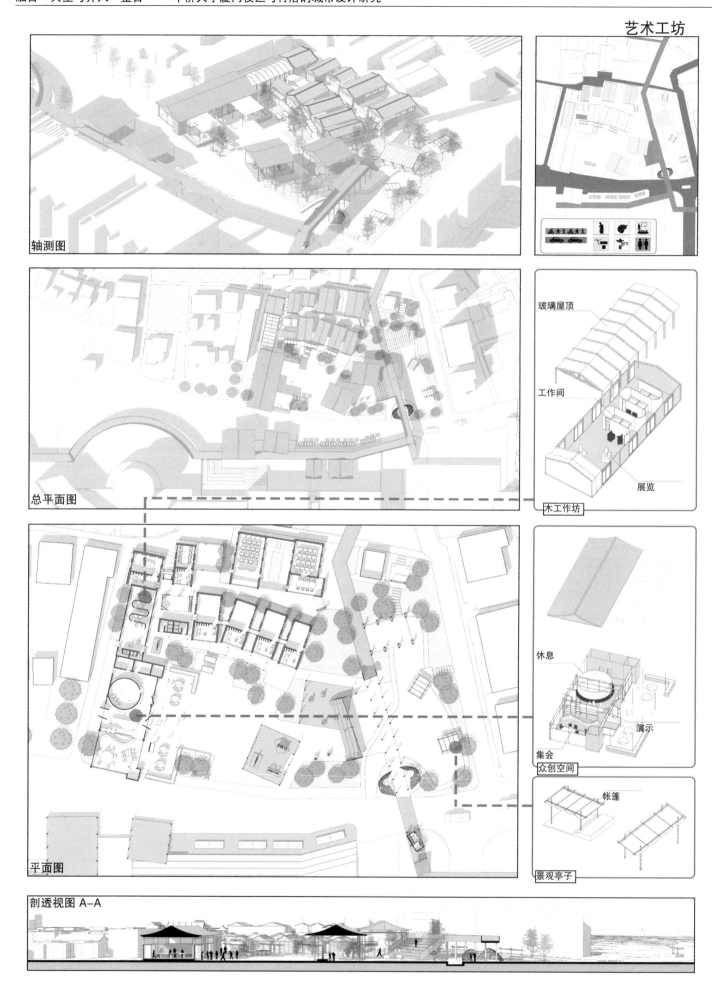

下篇：介入与整合 —— 基于校村共生的华侨文化保护与侨村环境整治方案设计

古榕公园

轴测图

总平面图

功能分布

城市家具

公园西侧

平面图

剖透视图 A-A

融合·共生与介入·整合 —— 华侨大学厦门校区与村落的城市设计研究

街环引侨
LOOPERS–A WALK TO CONNECT AND DEVELOP

指导老师：许懋彦
设计者：刘倩君 Germana Isacco

现状调研与问题分析

校村边界　道路结构　西头/西珩/南尾井

村中心　周边用地现状　下蔡/宅内/潘涂

建筑高度　保存价值　开放空间

概念与策略

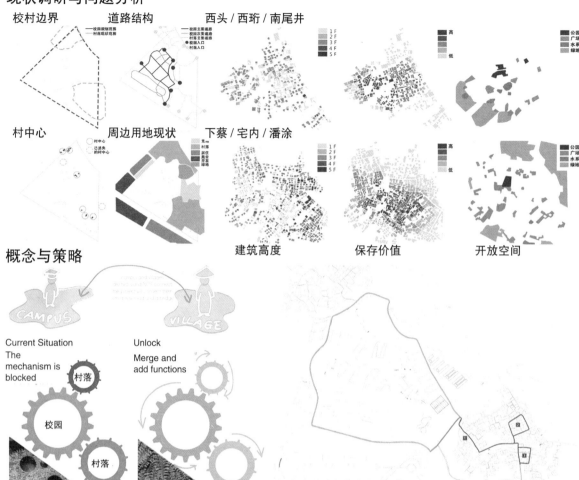

Current Situation
The mechanism is blocked

Unlock
Merge and add functions

村落
校园
村落

街道景观设计

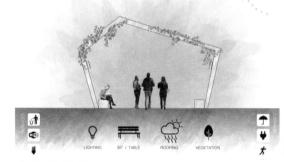

LIGHTING　SIT I TABLE　ROOFING　VEGETATION

中轴线设计

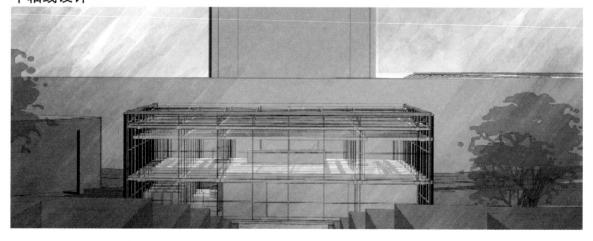

教师评语：

在校园与村落看似独立的个体中，彼此存在联动的可能。校园的拓展与村落街区复兴的互动成为本作品的关注点。以街道为线索的总体规划中，形成的校村互动模式，在功能混合上、空间织补上做出统筹协调。其中，规划视野从建筑到街区的拓展，带来了包括街道、建筑、环境的整体生态修复。弥补校村之间的时空错位，强化现有环境的特点。

下篇：介入与整合——基于校村共生的华侨文化保护与侨村环境整治方案设计

总平面图

1. 校园中轴线
2. 学生街
3. 桥村走廊
4. 改造民宿
5. 侨文化书院
6. 校园体育文化中心
7. 校园小亭
8. 新建教学楼
9. 民俗体验区
10. 休闲体育区
11. 宿舍区

规划结构分析

路网结构　　功能分区　　环线与节点　　新建建筑

融合·共生与介入·整合——华侨大学厦门校区与村落的城市设计研究

侨乡书院透视图

东立面局部图

组团与流线　开放空间　功能分布

首层平面图

下篇：介入与整合——基于校村共生的华侨文化保护与侨村环境整治方案设计

侨乡书院透视图

A-A 剖面图

流线与建筑类型　　功能分布　　开放空间

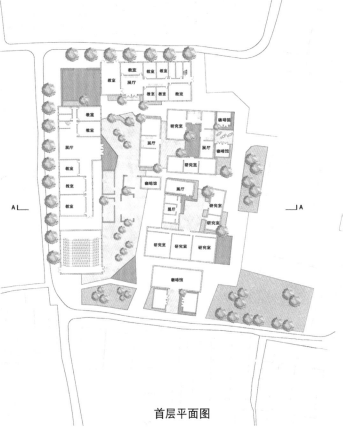

首层平面图

融合・共生与介入・整合 —— 华侨大学厦门校区与村落的城市设计研究

華僑大學 校村融合 之 華大八景
CONCEPTION RESEARCH

華大八景 / HQ EIGHT　　唐詩童　丁惟遲
華僑大學"介入與共生"整合計劃
指導教師　許懋彥 | 交圖日期 2017.06.18

華大八景　HUAQIAO EIGHT

指导老师：许懋彦
设计者：唐诗童　丁惟迟

教师评语：

城市之景，不仅是城市形象展示的窗口，更是浓缩城市人文地景的媒介。在华侨大学的校园内，作品试图摆脱实体物理空间的潜在限制，通过对特有景致的归纳提炼，描绘出隐藏在校村内部的精神特质，勾勒出属于华大的精神家园。在"华大八景"的选取、描绘和组织中，呈现出一张丰富的校园空间叙事图景。在共有的精神网络和实体标志的作用下，校与村将共同享人文价值，在求同存异下实现空间、功能与人群的融合。

建築圖底關係　　交通道路網　　八景及新游神線路結構　　輻射帶動區域　　建設后水系規劃

下篇：介入与整合——基于校村共生的华侨文化保护与侨村环境整治方案设计

華大八景之
概念研究
CONCEPTION RESEARCH

華 大 八 景 ／ HQ EIGHT　　唐詩童　丁惟遲
華僑大學"介入與共生"整合計劃

概 念 / CONCEPTS

华侨大学是由"中侨委"于1960年在周恩来总理直接关怀下创办的中央部属高校，有泉州、厦门两个校区。厦门校区选址于原兑山村之上，至今仍需处理新校区与旧庄之间的关系。本着保护原址文化、发挥华侨特色的原则，学校采取与村庄"校村共生"的理念。本设计尝试将古老的村落与高度现代化的校区，在"各为所用，互不干扰"的前提下，以一种更巧妙的方式进行缝合。初访场地之后，我们有感于校中村所蕴含的深厚乡村野趣的潜力。在当今飞速发展的城市里，如同几千年以前一样，乡村的引人遐想的美好使之再一次成了当代受教育者的精神故土。我们引用五首古诗文来阐释我们的乡村印象。

　　山重水复疑无路，柳暗花明又一村。——陆 游《游山西村》
　　雨里鸡鸣一两家，竹溪村路板桥斜。——王 建《雨过山村》
　　开轩面场圃，把酒话桑麻。——孟浩然《过故人庄》
　　阡陌交通，鸡犬相闻。——陶渊明《桃花源记》
　　千门万户曈曈日，总把新桃换旧符。——王安石《元 日》

我们认为，最不重要的并非其形态、功能，或是历史文化记忆，对于校园来说，最有意义的是这些乡村印象，我们希望在校园的设计中重现这些乡村特质，将乡村的诗意气质融入校园生活，让乡村的文人气质融入校园文化。

由于短期内的大拆大建不可行，村落应该是呈渐进式开发的格局被逐渐吸纳。脏乱差的环境势必会对使用造成影响。但我们相信，乡村的负外部性可以通过以点带面的设计被转化吃。我们从原址的形态、特征、以及村民的生活习惯中提取了我们认为最有意境与特色的八个点，希望通过这八个介入控制点来逐步激发村落活力；同时我们结合原址非常重要的游神路线，稍加改动，使之贯穿整个校园，不仅连接各个重要的宗祠，也将八景串联，在原来平庸直叙的校园规划中，加入了丰富的故事性和情节性，从而带动校村相融合。

八景的开放自由布局，给学校提供了尺度上的控制点，并辐射着周围组团的功能与使用，并能够在更高的文化维度上结合地段中蕴含的华侨文化，给在校学生提供接近、学习侨文化的契机，体现了华侨大学"会通中外、并育德才"的校训理念，并能够让在校师生、观光游客充分体会到学校丰富的自然景观和深刻的人文内涵。

Huaqiao University (HQU) was established by the Chinese central government in 1960 in Quanzhou, a coastal city in Fujian Province in the Southeast of China, for the education of overseas Chinese. With the campus combining with the villages, we decided to find the protential in 5 Chinese ancient poems to remaster the atmosphere in village.

A Visit to a Village West of the Mountains by Lu You, *The Raining Village* by Wang Jian, *Stopping by A Friend's Farm-House* by Meng Haoran, *The Peach Colony* by Tao Yuanming, and *New Year's Day* by Wang Anshi.

The five poems from famous poets of ancient Chinese depict a rustic charm of countryside that has been yearning by most of the people. We have tried to use these poems to guide our design. Integrating modern life, we intend to re-depict a poetic life in university that has shown in the mind of modern students. Such life has been gradually inspired by the control point of university. This design focuses on the "Eight Sights of Huaqiao University" and chooses eight most characteristic natural and cultural landscapes in schools and villages. The eight landscapes are independent, but are combined with the school's "new tourism route" which lead to the eight landscapes of Huaqiao University, Xiamen Campus. The author adds rich stories and scenes to the original unemphatic school planning.

By using the design method that combines with a plot, the author shows the rural scenery that cannot be seen in cities. Besides, by introducing the eight landscapes, a peculiar sight is displayed, in which schools and villages coexist, modern time and ancient elements occur together and neatness and rustic charm enhance each other's beauty. The open, free layout of the "Eight Sights of Huaqiao University" offer control points to school in terms of dimension. Besides, they influence the function and use of the surrounding groups. Integrating the culture of overseas Chinese in certain sections and in higher cultural range, the "eight sights of Huaqiao University" offer chances to students for getting in touch and studying the culture of overseas Chinese. Also, they reflect such school motto of Huaqiao University as "understanding both Chinese and foreign culture, cultivate students of education ability and political integrity". All these are aimed at allowing the teachers and students of the university as well as the tourists to fully appreciate the rich natural landscape and profound cultural connotation of the university.

研 究 / RESEARCH

宗祠文化

華僑大學

周邊村落

游神信仰

自然景觀

得出意象

標 識 / ICONS

田畝歸耕	北渚煙霞	花溪問道	翳影榕風
The original cultivated land of Xiheng Village will be reserved for the construction of a garden research center and experimental field in its original site, the ancient house area will be transformed into a farming culture museum.	With a pond the central landscape, there are teachers' canteen, tea room, exchange activity garden and other facilities built with ancient designs.	The river will be drained into the pool through the stream formed by height, and then it can be connected to the wetland system. The building is located at the intersection of the school's axis.	With birds singing in trees shades everywhere, the park is connected to the surrounding buildings through th paths along cliffs and platforms between trees, which form a restaurant in the old trees.

曲筵齋醮	壽捧兌丘	西洲晚棹	神遊泮宮
The original tour lines is cleaned up to form a small square. The old building will be changed to a school history museum and overseas Chinese culture research center. With houses and square stage staggered, it will become an unique place of interests.	A blanking tea pavilion on the other side is built on the axis of the core Lee Temple, the stone tablet behind it tells us the village's history.	The river will be drained into the pool through the stream formed by height, and then it can be connected to the wetland system. The building is located at the intersection of the school's axis.	With birds singing in trees shades everywhere, the park is connected to the surrounding buildings through th paths along cliffs and platforms between trees, which form a restaurant in the old trees.

游 神 / DEITIES PARADE

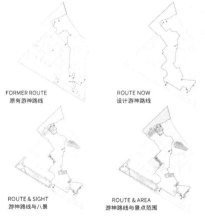

FORMER ROUTE　原有游神路線
ROUTE NOW　設計游神路線
ROUTE & SIGHT　游神路線与八景
ROUTE & AREA　游神路線与景點範圍

水 文 / WATER SYSTEM

CONTOUR MAP　建設前等高線
BEFORE　建設前校園水系
AFTER　建設前校園水系
FORMER PLAN　規劃校園水系
PLAN NOW　設計校園水系

兌山引水
景觀公園
階梯水系
碎水緩沖

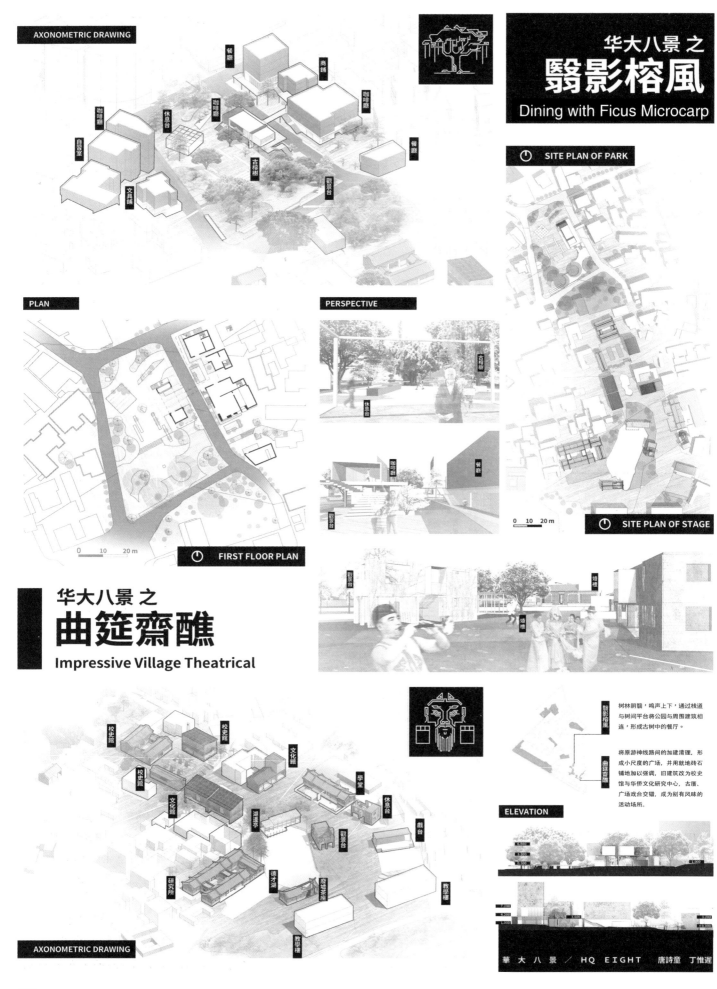

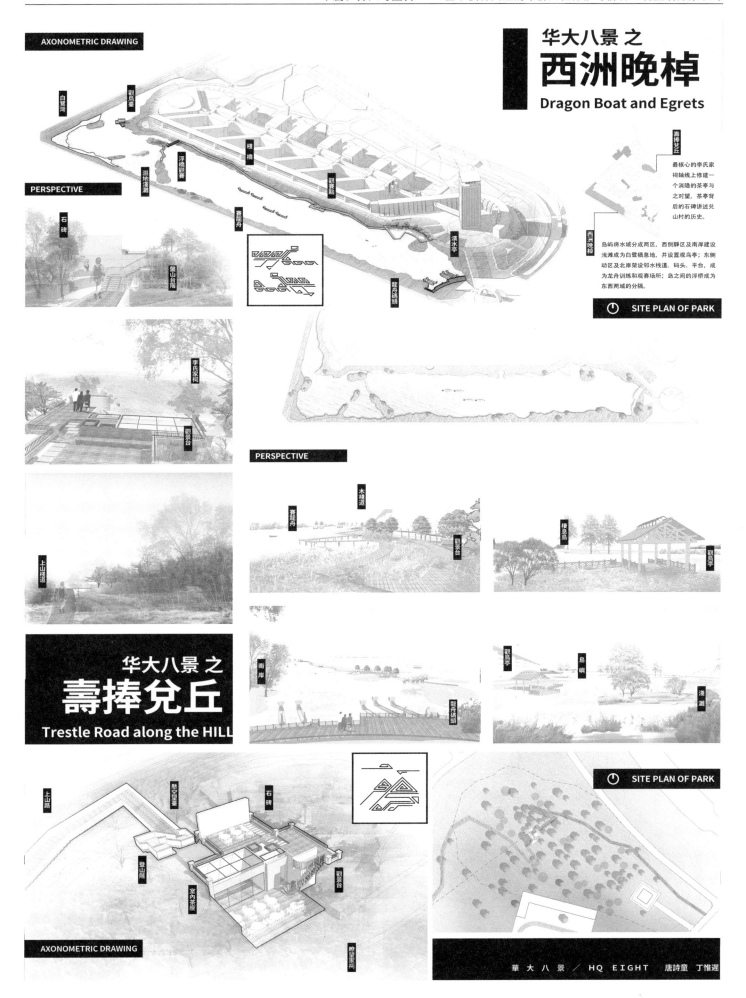

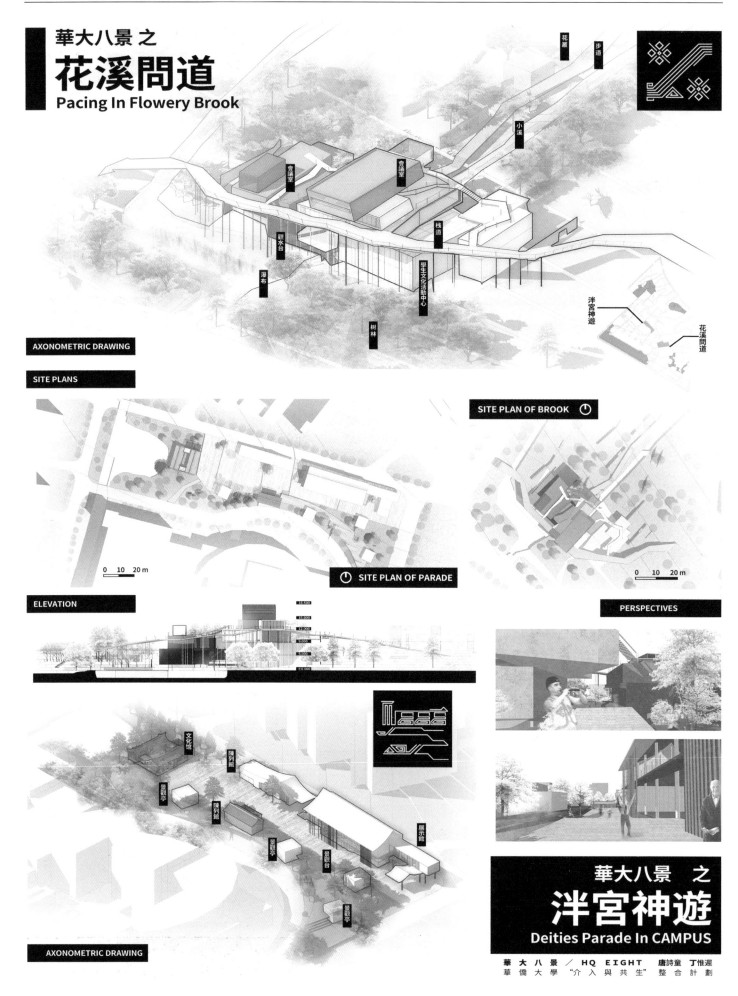

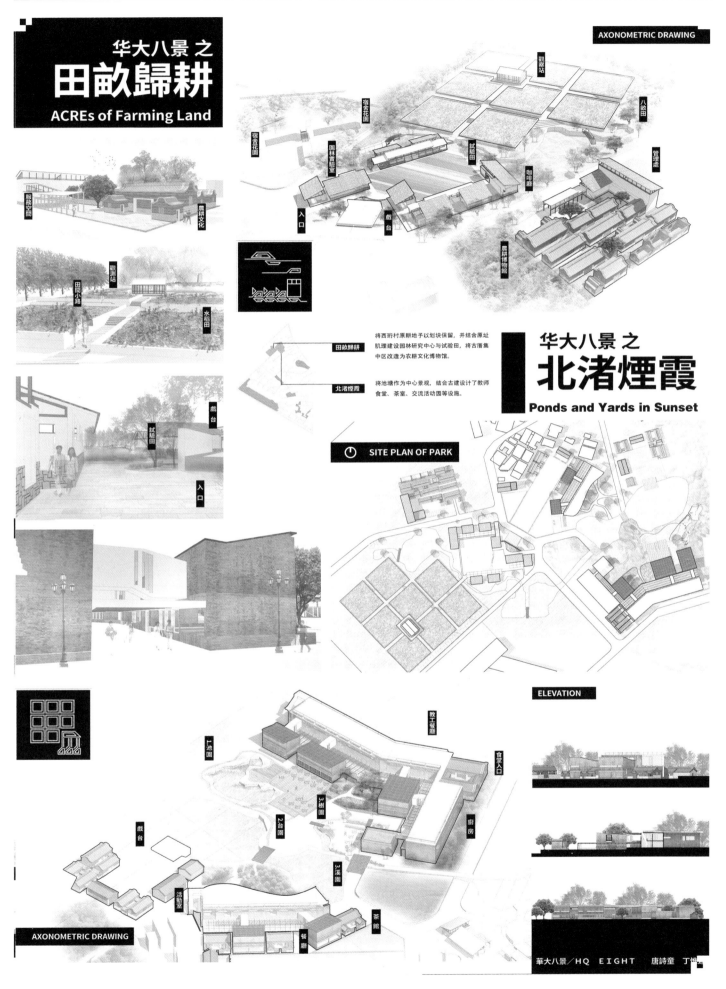

下篇：介入与整合 —— 基于校村共生的华侨文化保护与侨村环境整治方案设计

融合·共生与介入·整合——华侨大学厦门校区与村落的城市设计研究

Available space latent in the village
Utilization of space resources of village and transfer of university function
Spatial Sustainable Development of HQU's Xiamen Campus Project

AVAILABLE SPACE LATENT IN THE VILLAGE

指导老师：古谷诚章　课程助教：山田浩史
设计者：陈佳慧　斋藤隼　佐藤拓海　陈麟　陈路　王薪鹏

Analysis of Cultural Lines (Road, Wall, Water)
- Road Lines: street
- Boundary: walls and districts
- Tiny Lakes: ponds
- Squares by Temples: Historical Cores
- Temples and Stages: Historical Cores
- Old Trees: Historical Cores

Analysis of Cultural Spots
- Green Spots: Green Cores; so-called Main Cores
- Green Spots: ; so-called Sub Spaces
- Tiny Lakes: Ponds
- Squares by Temples: Historical Cores
- Temples and Stages: Historical Cores
- Old Trees: Historical CORES

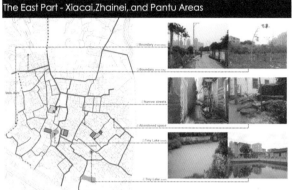
The East Part - Xiacai, Zhainei, and Pantu Areas

The East Part - Xiacai, Zhainei, and Pantu Areas

The North part - Xiheng, Nanweijing, and Zucuohou Areas

The North Part - Xiheng, Xitou, Nanweijing, and Zucuohou Areas

The Space with potential

Wall　　Vacant house　　Historical building　　Park

The Space that began to be utilized

教师评语：
　　该方案引入了"木造租房recipe"这一日本现在流行的空屋再生的手法。尺度上，多落在建筑立面、室内、家具的尺度；手法上，通过分析整理住宅本身的多个细节来进行针对性设计；对象上，选取了临接校区的住宅和功能用地；可供用途上，既可用作商业、公用等。但很明显，首先是日本的手法在中国的适用性如何；同时，这种针对细节的设计需要详细调查才能具有针对性，这不是简单泛泛的概况调查就足够了。

Gallery　　Rest area　　Barn　　Vegetable garden

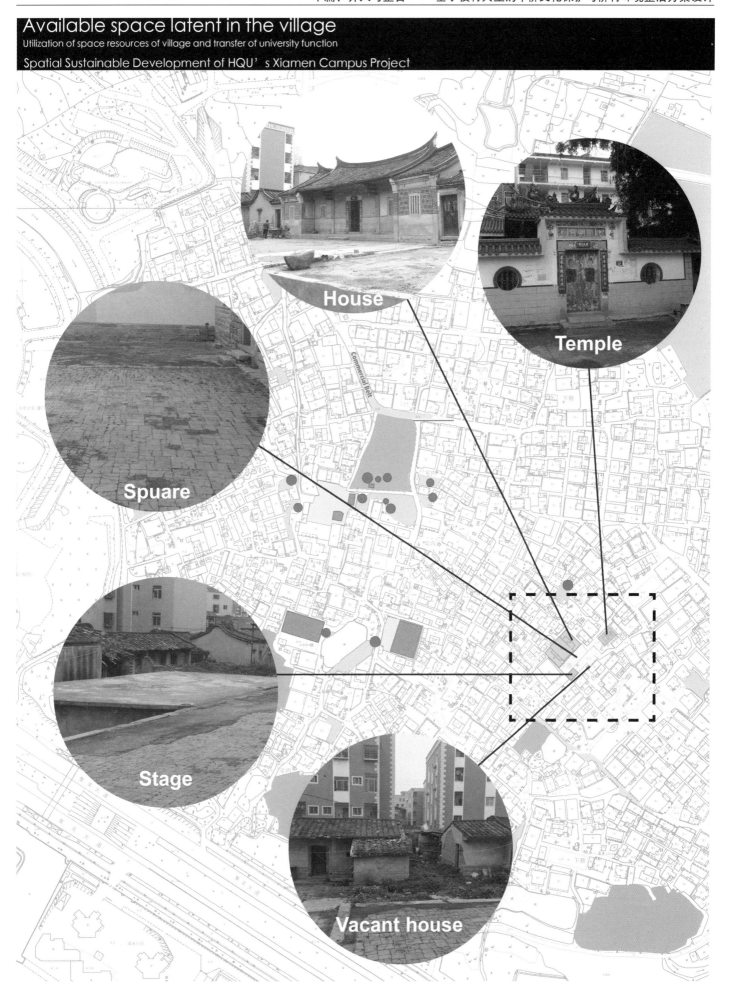

Available space latent in the village
Utilization of space resources of village and transfer of university function
Spatial Sustainable Development of HQU's Xiamen Campus Project

At night, the stars are beautiful.....
Balco

When you want to be alone, you can go up the floor!
Roof

The appropriate height make new communications and uses
Stairs

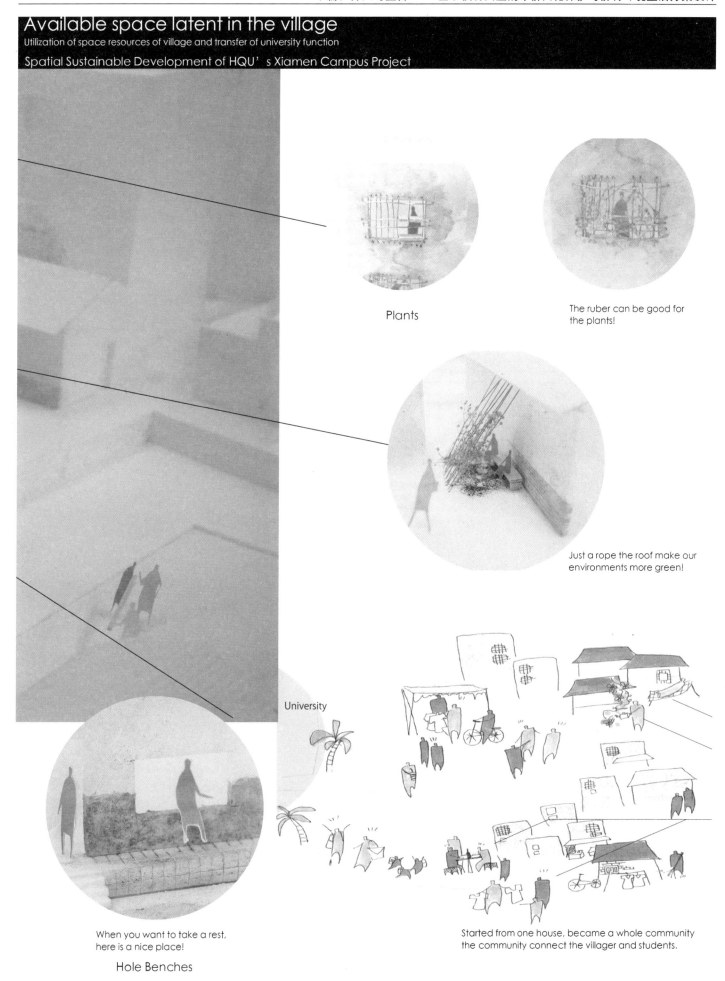

重庆大学
CHONGQING UNIVERSITY

目录 CATALOG

戏致入微　张叶菁　汪天琦　邱鞠
Intervention as Immersion

朴垣织侨　曹雅涵　何格　张宇
Patch the Texture & Weave the Culture

共聚·侨间　邵瑾　彭雨轩　梁晨
Reunion in the Qiao Culture

侨路蔓蔓　左云豪　芮万里　王创懿
Overseas Road Slowly Spread

下篇：介入与整合 —— 基于校村共生的华侨文化保护与侨村环境整治方案设计

学生团队 Student Team

何格　曹雅涵　芮万里　邵瑾　王创懿　张叶青
汪天琦　左云豪　张宇　彭雨轩　邱鞠　梁晨

融合·共生与介入·整合 —— 华侨大学厦门校区与村落的城市设计研究

侨路蔓漫
OVERSEAS ROAD SLOWLY SPREAD

指导老师：邓蜀阳 阎波
设计者：左云豪 芮万里 王创懿

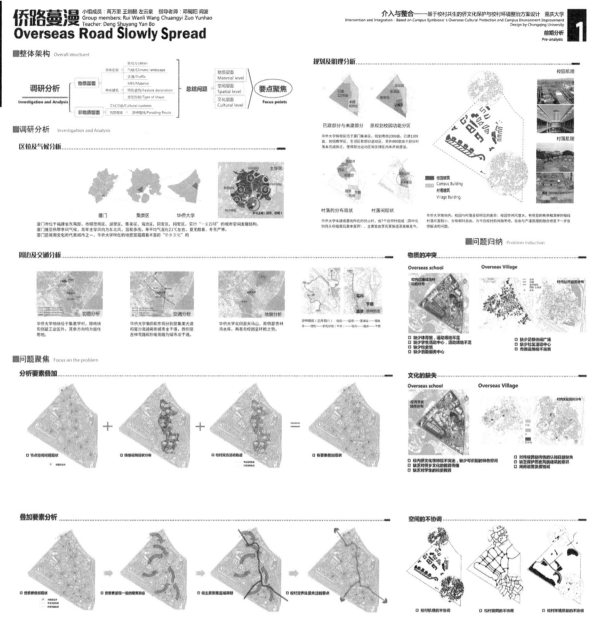

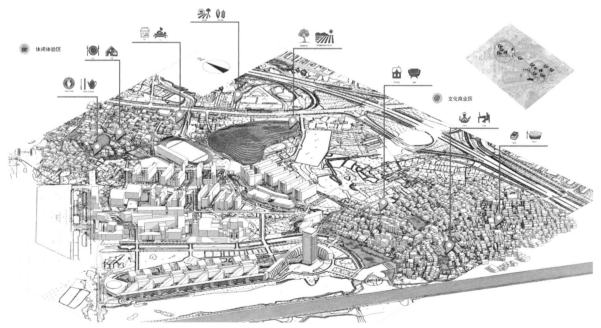

教师评语：
作品从村落现状入手，分析出场地内现有侨文化缺乏、居民对传统民俗的认知缺失以及校村环境不协调等校村之间问题，通过织补校村路网、公共空间、节点整治等方法，把承载侨乡的文化要素通过公共中心、游神步道所形成的侨文化长廊展现出来，并串接整个校村，实现校村共融。

下篇：介入与整合 —— 基于校村共生的华侨文化保护与侨村环境整治方案设计

侨路蔓漫
Overseas Road Slowly Spread

小组成员：芮万里 王创懿 左云豪
Group members: Rui Wanli Wang Chuangyi Zuo Yunhao
指导老师：邓蜀阳 闫波
Teacher: Deng Shuyang Yan Bo

介入与整合——基于校村共生的侨文化保护与校村环境整治方案设计 重庆大学
Intervention and Integration — Based on Campus Symbiosis's Overseas Cultural Protection and Campus Environment Improvement Design by Chongqing University

文化长廊 2
Culture Gallery

■ 文化长廊总平面图
Site Plan of culture gallery

■ 概念演绎
Design Concept

校村隔阂　文化藤蔓　蔓延渗透

■ 文化长廊形成
The Culture Gallery formation

1. 华侨大学 Hua Qiao University
2. 村落区域 Village area
3. 校村边界确定 boundary
4. 校村路网梳理 Road Network
5. 游神步道叠加 Youshen Road
6. "影响域" Influence domain
7. 校园公共空间梳理 Campus public space

9. 形成文化长廊 Cultural corridor

■ 规划结构分析
The Analysis Of The Culture Gallery

道路结构 path
功能分区 function
空间节点 Public space
单体位置 New building

■ 文化长廊结构分析
The Structure Of The Culture Gallery

侨村民宿
手工艺作坊
侨乡民俗馆
文化长廊
兑山村
华侨大学校园

■ 文化长廊定位
The Definition Of The Culture Gallery

休闲体验区
文化商业区

■ 校村交界处界面
The Boundary Between The School And Village

校园主入口 | 校村广场 | 村落 | 校园运动场馆 | 校村广场 | 村落

融合·共生与介入·整合 —— 华侨大学厦门校区与村落的城市设计研究

侨路蔓漫
Overseas Road Slowly Spread

小组成员：芮万里 王创懿 左云豪　指导老师：邓蜀阳 阎波
Group members: Rui Wanli Wang Chuangyi Zuo Yunhao
Teacher: Deng Shuyang Yan Bo

介入与整合——基于校村共生的侨文化保护与校村环境整治方案设计　重庆大学
Intervention and Integration - Based on Campus Symbiosis's Overseas Cultural Protection and Campus Environment Improvement
Design by Chongqing University

整体策略 3
Holistic strategy

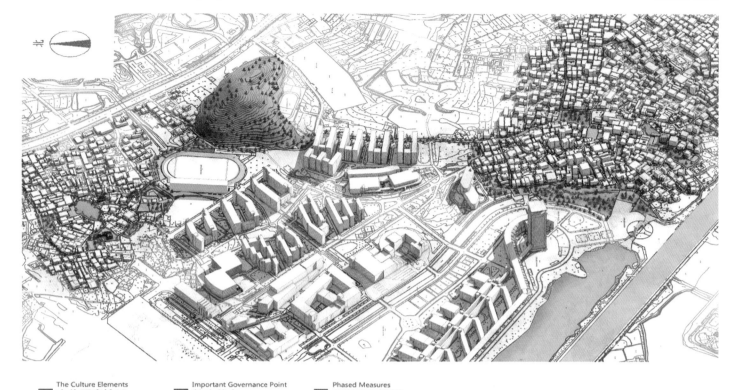

文化要素梳理
The Culture Elements

整治节点
Important Governance Point

分期治理措施
Phased Measures

近期　中期　远期

对整个文化长廊上的文化要素进行梳理，例如荷堂庙庵、古榕公园、池塘以及旁神路线。这些文化要素是整个村落的文化载体和文化寄托，也是连接校村最强有力的结合剂。

整个文化长廊的总体治理分为近期，中期，远期三个部分。
近期以文化要素的提取和初步环境治理为主；
中期以加强文化认同为主，加强文化纽带的连接力度；
远期则主要以校村文化活动为主，形成介入和渗透。

专项措施
Details Measures

路灯
Street Lamp

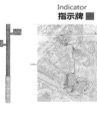
指示牌
Indicator

古榕公园
Park

场地现状　改造整治后

围墙
Wall

场地现状　改造整治后

池塘
Pond

场地现状

改造整治后

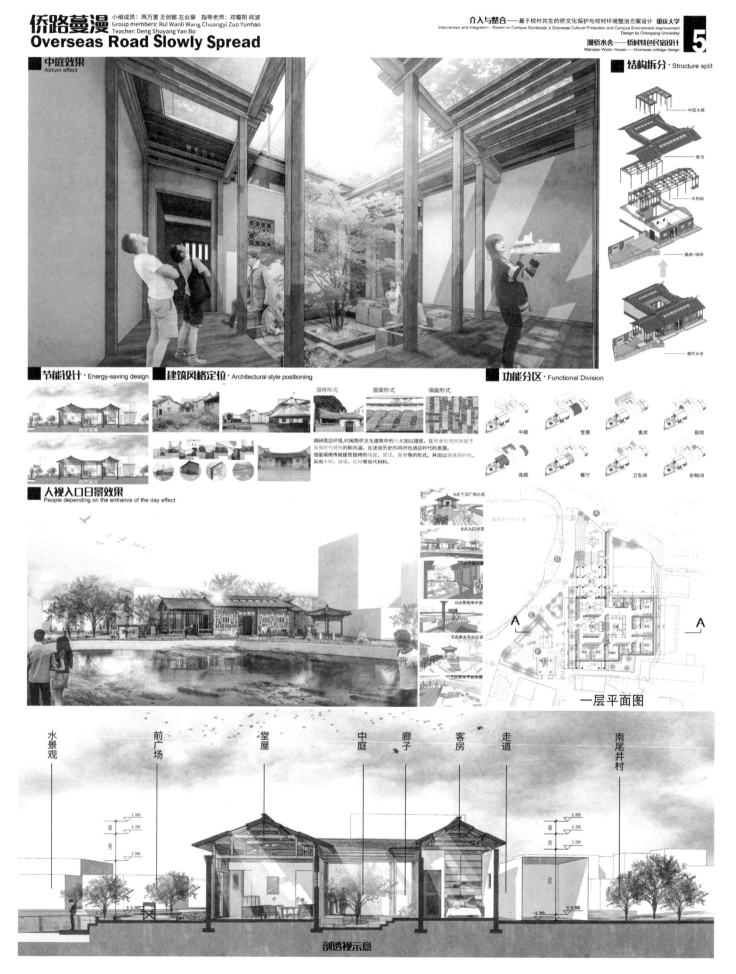

下篇：介入与整合 —— 基于校村共生的华侨文化保护与侨村环境整治方案设计

侨路蔓漫
Overseas Road Slowly Spread

小组成员：芮万里 王创懿 左云豪　指导老师：邓蜀阳 阎波
Group members: Rui Wanli Wang Chuangyi Zuo Yunhao
Teacher: Deng Shuyang Yan Bo

介入与整合——基于校村共生的侨文化保护与校村环境整治方案设计　重庆大学
Intervention and Integration - Based on Campus Symbiosis's Overseas Cultural Protection and Campus Environment Improvement Design by Chongqing University

闽语侨乡——侨乡民俗馆
Min Yu Qiao Xiang——Folk Museum design

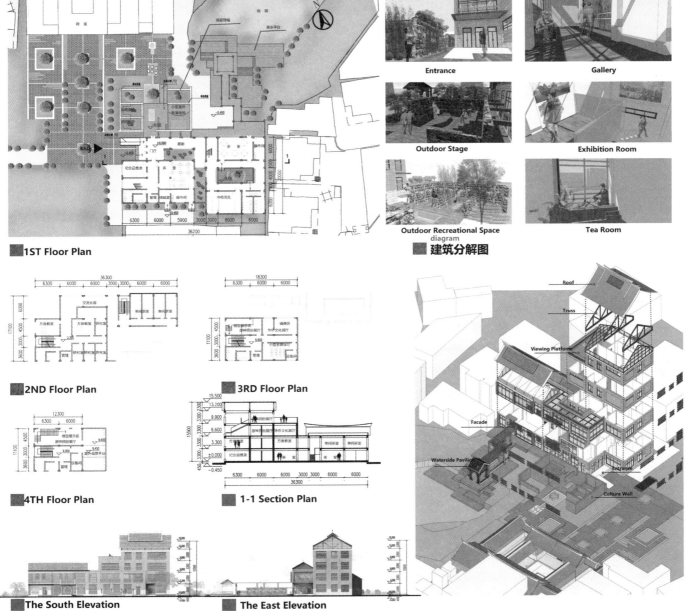

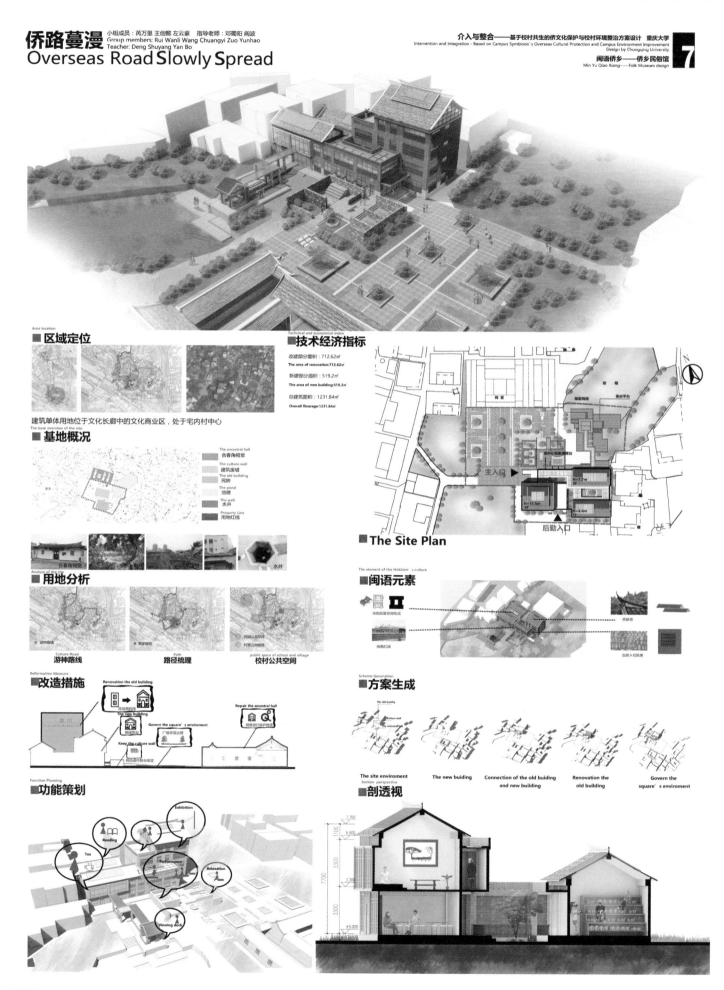

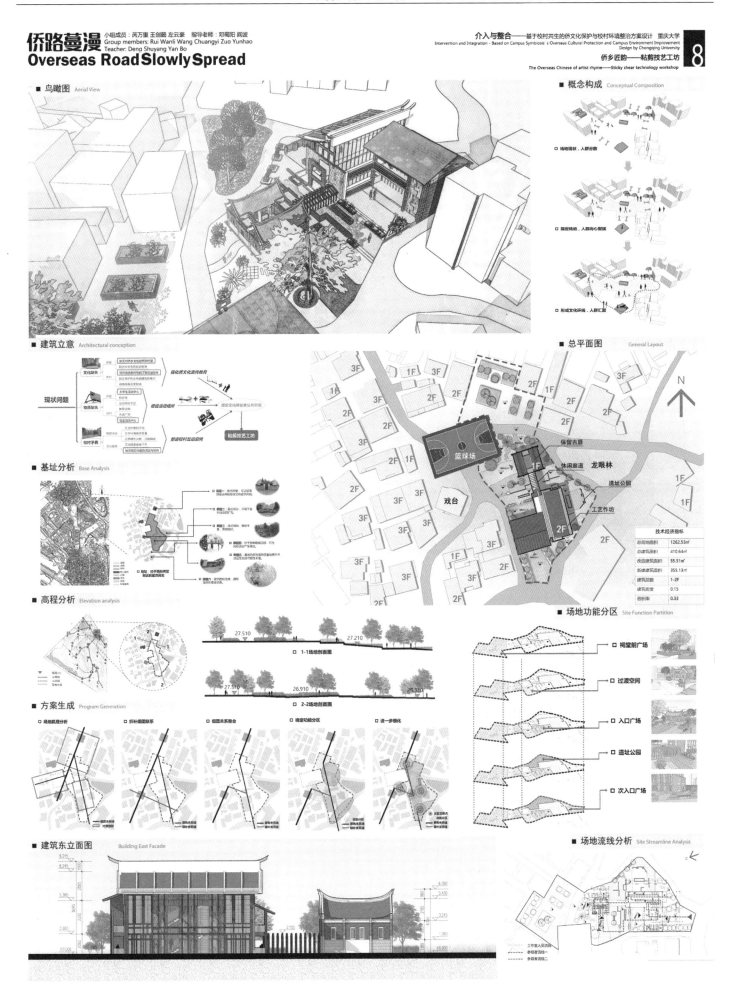

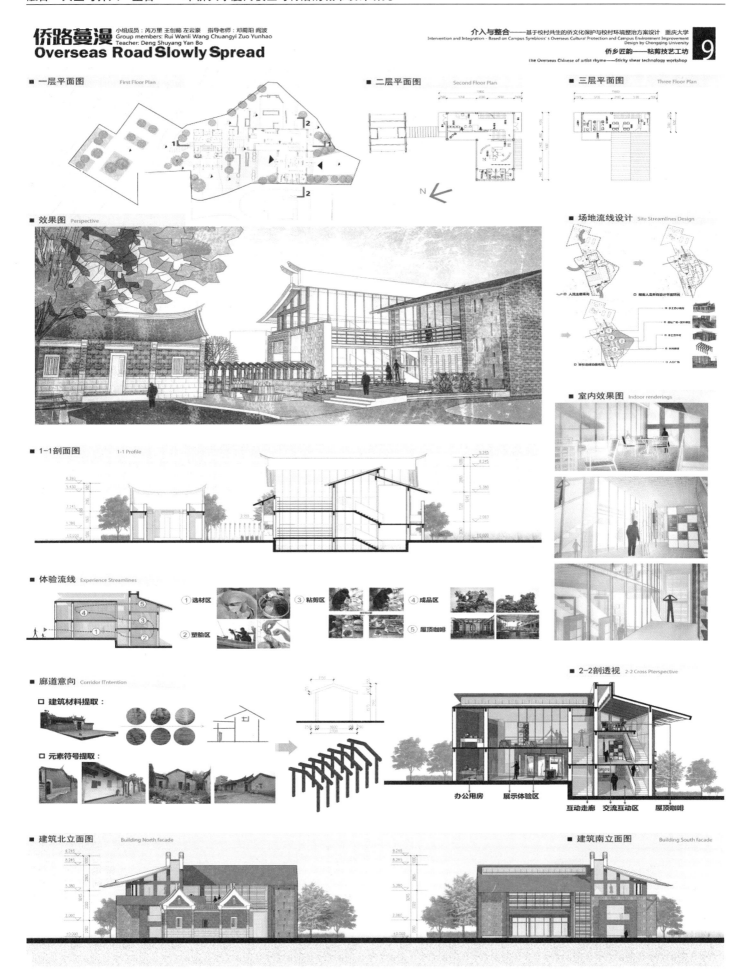

下篇：介入与整合 —— 基于校村共生的华侨文化保护与侨村环境整治方案设计

共聚·侨间
REUNION IN THE QIAO CULTURE

介入与整合：基于校村共生的华侨文化保护与侨村环境整治方案设计

华侨大学厦门校区&兑山村——前期分析 01

指导老师：邓蜀阳 阎波
设计者：邵瑾 彭雨轩 梁晨

区位背景 Location background

本次设计的研究范围是中国福建省厦门市华侨大学厦门校区，该校区位于集美文教区、集美大道北侧，背靠天马山，面对杏林湖，现已完成大部分规划。校园教学区、生活区基本建设完成。建设基地原属于兑山村，集中于下寮、潘涂和宅内三甘。

The research scope of this design is China's fujian province xiamen xiamen overseas Chinese university campus, the school district is located in jimei culture and education district, north of jimei avenue, back relies on mashan, in the face of almond shaped lake, have been completed most of the planning. The construction of campus teaching areas, living area basic finish. Under the construction base belonging to the village, to focus on CAI, pan coating, and three clubs within the house.

历史演变 Historical evolution

- 2003 华侨大学建校以前 The overseas Chinese university's ago
- 2005 学校与村落同时发展 Schools and villages develop simultaneously
- 2009 学校扩张与村落发展停滞 School expansion and village development stalled
- 2013 学校进一步压缩村庄空间 The school further compressed the village space
- 2017 用地紧张引起的校村对立 The land tension causes the school village opposite

公共空间 public space

- 戏台 Stage — 现状：现状较差，缺乏管理，使用率低 Status: poor status, lack of management; low utilization rate
- 古树 Old trees — 现状：分布零散，生存状况岌岌可危 Status: scattered and precarious

- 宗祠 temple — 现状：利用率较低，祠堂活力不足 Status: low utilization rate, lack of vitality of the temple
- 街巷 Street — 现状：主路拥堵，次路狭窄，缺乏整治 Status: the main road is congested, the secondary road is narrow, lack of

校村结构 School village structure

一、道路系统
道路系统差异明显，缺乏有效的沟通和衔接，缺失导向性和合理规划。
First, the road system — The difference of road system is obvious, lacking effective communication and cohesion, lack of guidance and rational planning.

二、公共空间
学校较好，村落较差，两者应结合各自特点实现优势互补。
Second, the public Spaces — The school is better, the village is poorer, the two should combine each characteristic to realize the superiority complementary.

- 盾埕 House courtyard — 现状：缺少铺装和绿化，缺乏打理 Status: lack of pavement and greening, lack of care
- 池塘 pond — 现状：绿藻异味，垃圾污染问题严重 Status: green algae odor, garbage pollution problem serious

三、校村结构
村落与学校基本呈现对立姿态，以后代作为两者的隐性分隔。
Third, The school village structure — Villages and schools are basically opposites, and the street as a recessive separation of the two.

四、人员对比
校内为老师和学生，村中为村民和工人，双方有交叉但融合度较低。
Fourth, Researchers compared — The campus is for teachers and students, the village is for the villagers and the workers, the two sides have crossed but the harmony is low.

- 水井 well — 现状：大部分荒废，水质和周边环境较差 Status: most of the waste water and surrounding environment are poor
- 空地 space — 现状：被拆迁后形成荒地，缺乏有效管理 Status: after being demolished to form wasteland, lack of effective man

概念解析 Conceptual analysis

三大策略 Three strategies

   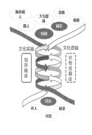

八大分区 Eight divisions

- 挖掘文化源点 Dig the cultural source
- 激活文化源点 Activate the cultural source
- 打通外围环道 Open the peripheral ring road
- 拓宽南北道路 Broaden the north and south roads
- 打通东西道路 Get through the east and west road
- 塑造文化场域 Shaping cultural areas

阶段一 Stage One / 阶段二 Stage Two / 阶段三 Stage Three / 阶段四 Stage Four

教师评语：
华侨大学新校区内校村共存，虽然两者空间结构不协调，但共处在一个空间之中。作品紧扣两者之间的"共聚"特点，把传统文化复兴、村落环境修复、民俗活动植入作为"共聚"的触媒，制定出激活侨文化的三大策略，把整个校园空间做了八大分区，并拟定了校村整治行动的近、中、远期行动建议，实现校村在空间网络、文化网络上的共生。

融合·共生与介入·整合 —— 华侨大学厦门校区与村落的城市设计研究

共聚·侨间

介入与整合：基于校村共生的华侨文化保护与侨村环境整治方案设计

Regional Overall and System Design 区域系统设计 02

Intervention and Integration: Preservation and Revitalization Design for the Traditional Village-within-the Campus in Huaqiao University

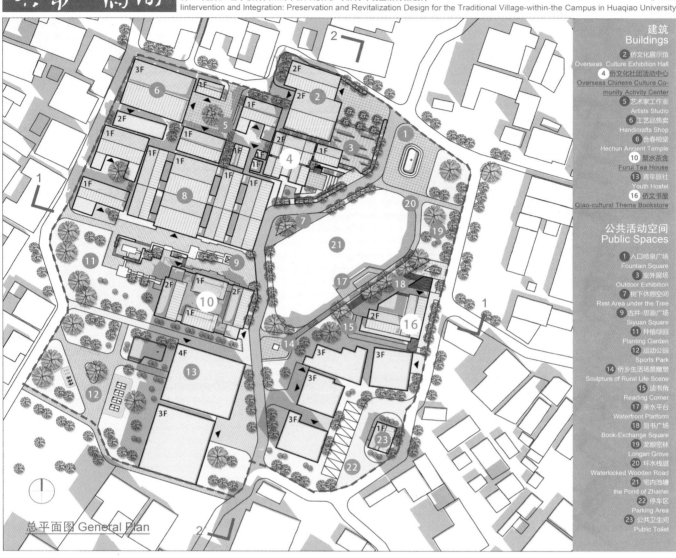

总平面图 General Plan

建筑 Buildings
- ② 侨文化展示馆 Overseas Culture Exhibition Hall
- ④ 侨文化社团活动中心 Overseas Chinese Culture Community Activity Center
- ⑤ 艺术家工作室 Artists Studio
- ⑥ 工艺品售卖 Handicrafts Shop
- ⑧ 合春榕堂 Hechun Ancient Temple
- ⑩ 聚水茶舍 Furui Tea House
- ⑬ 青年旅社 Youth Hostel
- ⑯ 侨文书屋 Qiao-cultural Theme Bookstore

公共活动空间 Public Spaces
- ① 入口喷泉广场 Fountain Square
- ③ 室外展场 Outdoor Exhibition
- ⑦ 树下休憩空间 Rest Area under the Tree
- ⑨ 古井·恩源广场 Siyuan Square
- ⑪ 种植绿园 Planting Garden
- ⑫ 运动公园 Sports Park
- ⑭ 侨乡生活场景雕塑 Sculpture of Rural Life Scene
- ⑮ 读书角 Reading Corner
- ⑰ 亲水平台 Waterfront Platform
- ⑱ 易书广场 Book-Exchange Square
- ⑲ 龙眼密林 Longan Grove
- ⑳ 环水栈道 Waterlocked Wooden Road
- ㉑ 宅内池塘 the Pond of Zhainei
- ㉒ 停车区 Parking Area
- ㉓ 公共卫生间 Public Toilet

重要节点透视 Perspective of Important Nodes

古井·恩源广场 Siyuan Square

侨乡生活场景雕塑 Sculpture of Rural Life Scene

亲水平台 Waterfront Platform

各系统设计前后对比 Comparison of System

建筑肌理 Architectural Texture
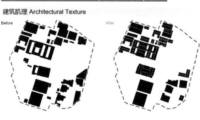

交通 Transportation
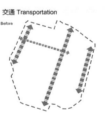
机动车道 Roadway　人行道 Pavement

建筑性质及状况 Architectural Attributes and Situation

公共 Public　私人 Private　废墟 Ruins　其他 Others

建筑风貌形式 Architectural Style

传统 Traditional　现代 Modern

广场及公共院落 Squares and Public Yards
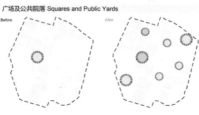

景观 Landscape

共聚·侨间
介入与整合：基于校村共生的华侨文化保护与侨村环境整治方案设计
环境整治策略及行动指引 03
Intervention and Integration: Preservation and Revitalization Design for the Traditional Village-within-the Campus in Huaqiao University

场地现状 Site Status

| 文化层面 Cultural level | 道路层面 Road level | 建筑层面 Building level | 环境层面 Environmental level | 绿化层面 Greening level | 铺装层面 Pavement level | 设施层面 Facility level |

整治策略 Remediation strategy

| 文化层面 Cultural level | 道路层面 Road level | 建筑层面 Building level | 照明层面 Lightning level | 绿化层面 Greening level | 铺装层面 Pavement level | 设施层面 Facility level |

行动指引 Action guidelines

	治理近期行动建议 Recent action recommendations		治理中期行动建议 Medium-term action recommendations
公共服务设施缺乏	1) 垃圾桶：根据人流与建筑密度，按需求配置数量合理的垃圾桶 2) 路灯：设置南北主要道路、东西道路、环湖道路路段的路灯 3) 健身设施：考虑不同年龄段人群需求，增设游乐休憩、健身设施	校村人际关系	1) 学生毕业季举办跳蚤市场，村民购置铺铺书籍 2) 村民为校方研究人员提供家书家史，校方为村民提供历史知识科普教育
场地环境污染	1) 整个场地垃圾分区分块管理，设置相应检查管理机构进行健康整治 2) 提高垃圾收运频率，减少垃圾停留时间	文化体育活动组织	1) 文化类：举办露天观影活动、读书交流活动 2) 体育类：举办校村联合趣味运动会、社区羽毛球赛
建筑缺乏维护	1) 政府提供补贴村民出资修缮保护祠堂与现存古屋 2) 校方对外开放吸引外来方对废墟进行改造功能置换	侨乡对外旅游开发	1) 组织主要旅游路线，完善相关产品打造、当地手工艺纪念品、历史知识宣传带 2) 校方对外招商，适当合理引入外来商户，结合古聚落进行传统文化新发展的改造
活动场地缺乏管理	1) 社区组织专人负责活动场地的环境卫生及设施维护 2) 引导村民、社会组织定点开展各类活动，丰富社区生活 3) 室内外活动场地行分时段管理，提高场地利用效率	中理远期行动建议	Forward action recommendations
		侨文化宣传组织	1) 建立兑山村侨文化宣传队，负责对清涂、宅内、南尾井等各个村的历史文化挖掘和宣传，培养大学生对游客进行侨乡旅游文化的讲解
公共空间不合理占用	1) 设置监控与相关管理人员，对不合理占用公共场地的车辆进行罚款处理，并收取停车费限制场地内车辆数量 2) 拆除私宅违建搭建，清理在公共场所堆放的杂物，疏通被院侨所封闭的原乡村落道路与场地	大学生志愿者组织	1) 扶持公益类学生社区组织，挖掘校方与村方的人力资源，优势互补，建造粘贴技艺工作室，周都文化研究、社区公共设施维护队等志愿者组织
		兑山社区文体组织	1) 发展文体类社区组织，利用改造后的活动场地，组建羽毛球、乒乓球等文体组织，丰富村民生活，加强校村交流，提高双方对校村的认同感与归属感

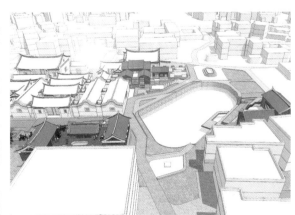

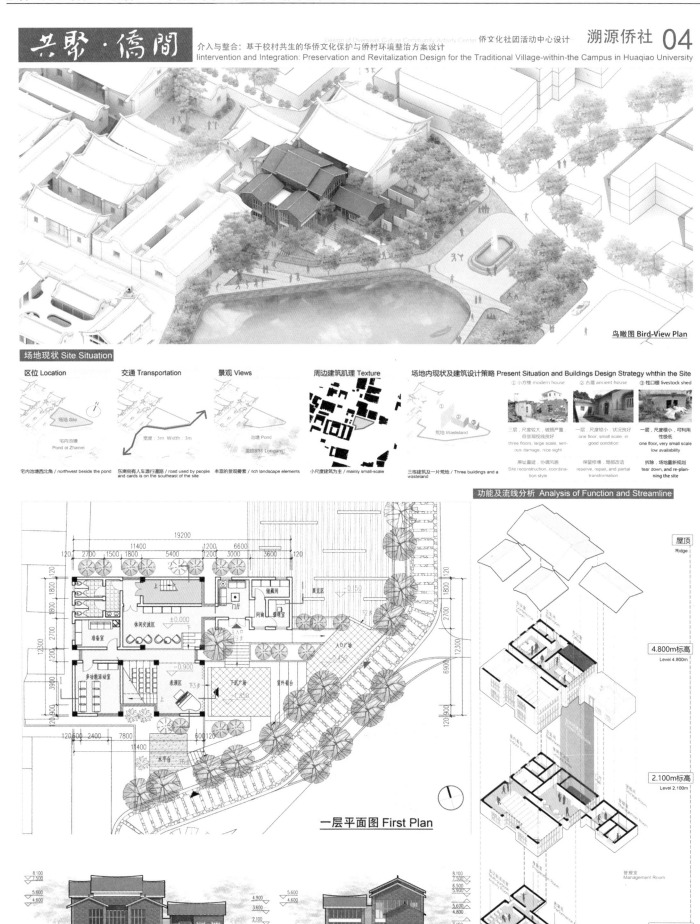

下篇：介入与整合 —— 基于校村共生的华侨文化保护与侨村环境整治方案设计

共聚·侨间
介入与整合：基于校村共生的华侨文化保护与侨村环境整治方案设计
侨文化社团活动中心设计——溯源侨社 05
Intervention and Integration: Preservation and Revitalization Design for the Traditional Village-within-the Campus in Huaqiao University

技术经济指标 Techno-Economic Indexes
- 总用地面积 Total Site Area: 410.32 m²
- 建筑密度 Intensity: 42%
- 建筑基底面积 Base Area: 174.23 m²
- 容积率 Plot Ratio: 0.70
- 总建筑面积 Total Floor Area: 287.54 m²

总平面图 General Plan

水边视角 View from the Pond

二层平面图 Second Plan

1-1 剖面图 1-1 Section Plan

2-2 剖面图 2-2 Section Plan

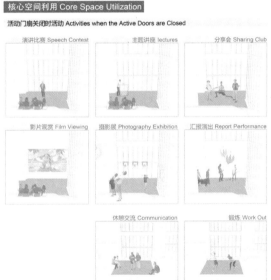
核心空间利用 Core Space Utilization

活动门扇关闭时活动 Activities when the Active Doors are Closed
- 演讲比赛 Speech Contest
- 主题讲座 Lectures
- 分享会 Sharing Club
- 影片观赏 Film Viewing
- 摄影展 Photography Exhibition
- 汇报演出 Report Performance
- 休憩交流 Communication
- 锻炼 Work Out

活动门扇开启时活动 Activities when the Active Doors are Opened
- 地方戏曲表演 Local Opera
- 互动体验 Interactive Experience

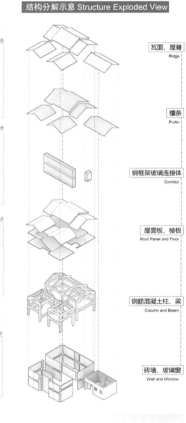
结构分解示意 Structure Exploded View
- 瓦面、屋脊 Ridge
- 檩条 Purlin
- 钢框架玻璃连接体 Corridor
- 屋面板、楼板 Roof Panel and Floor
- 钢筋混凝土柱、梁 Column and Beam
- 砖墙、玻璃窗 Wall and Window

Cross perspective of the Core Space When the Active Doors are Opened
活动门扇开启状态核心空间剖透视

183

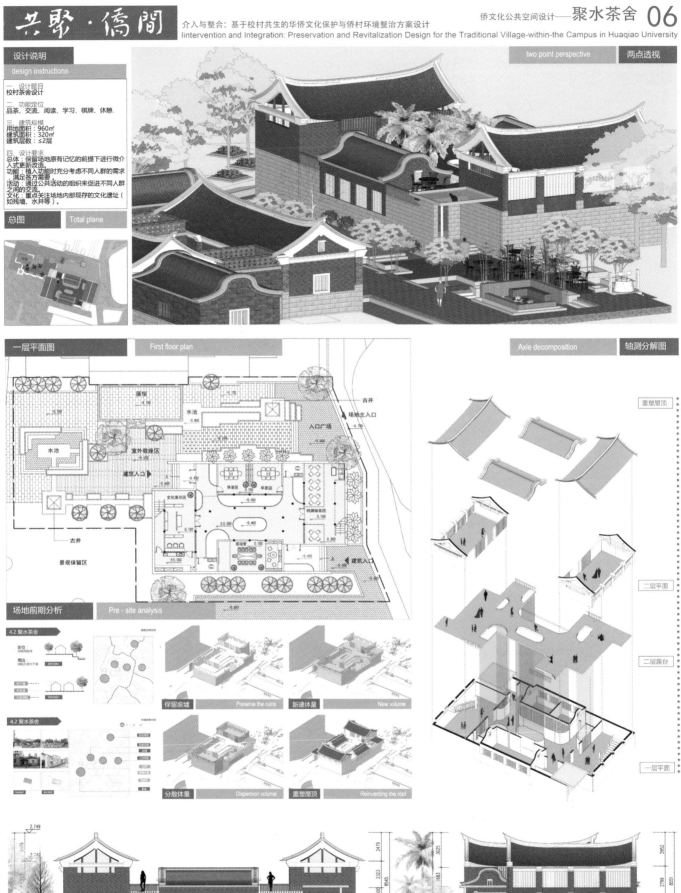

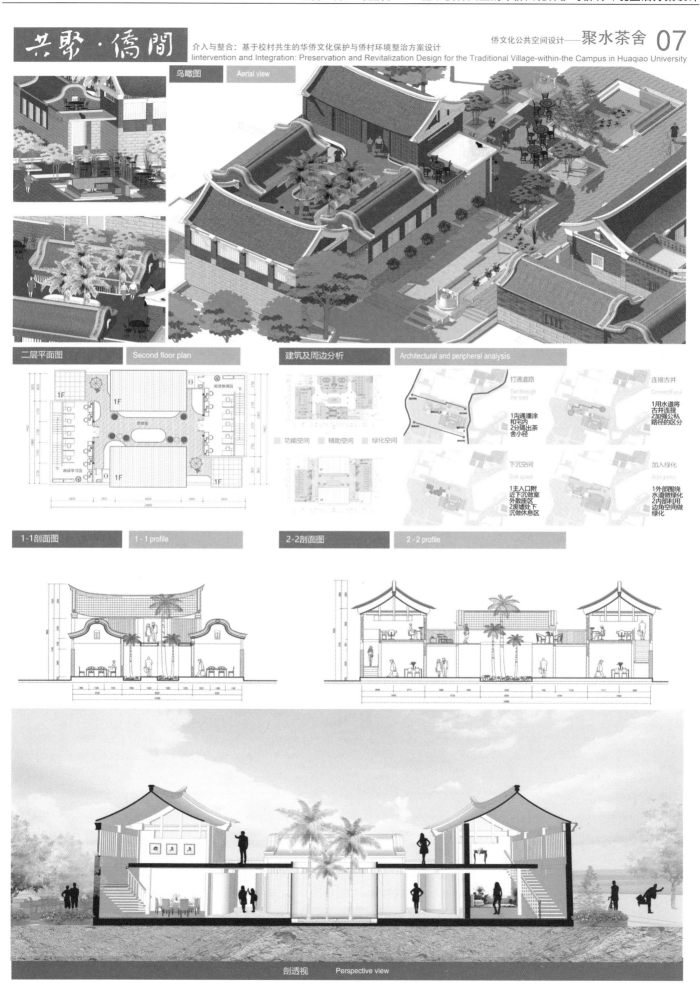

共聚·侨间

介入与整合：基于校村共生的华侨文化保护与侨村环境整治方案设计

侨文化主题微型图书馆设计——侨文书屋 08

Intervention and Integration: Preservation and Revitalization Design for the Traditional Village-within-the Campus in Huaqiao University

场地现状 Renderings

背景思考 Background thinking

无论是华侨大学校园还是兑山村，都没有一个供人专门了解学习侨文化的地点
校方师生、外来租客与原住村民缺乏交流沟通，尤其是文化意义上的
兑山村内多处公共空间缺乏设计引导，没有人气与活动，日益闭塞与没落

思考：
解决文化问题——以书作为切入点，微型图书馆的介入
解决交流问题——围绕图书馆活动，吸引不同人员的到来
解决场地问题——在维系池塘附近原有肌理的基础上适当改，最小负担最大利益

设计：
功能：侨文化——主题微型图书馆（建筑总面积250+平方米）
定位：面对网络电商与大型书店空隙 具有特色的小型独立书店
目标：保护宣传侨乡文化、促进校村双方交流、激活池塘失落空间

设计任务书 Design task

功能 function	面积 area	数量 number	备注 remarks
阅览 read	40	2	划分新书与旧书区
售卖 sale	50	1	划分纪念品区、尚书区、展览区等等
休闲 relax	30	1	设置咖啡设施
儿童 children	20	1	以矮墙划分空间便于家长看护
旧书架 stack	15	1	篮子一样便于寄送
办公 office	12	1	
卫生间 WC	12	2	每层一，可男女共用
入口广场	50	1	
易书广场	70	1	
读书角	30	1	结合原有龙眼树布置
休息平台	40	1	
亲水平台	40	1	

总平面图 General Layout

总平面图 General Plan

场地现状 Site Status

地块高差 Height difference
围墙阻隔 Barrier wall
违章建筑 Illegal construction
中心池塘 Center pond
龙眼树林 Longan woods

功能需求 Functions

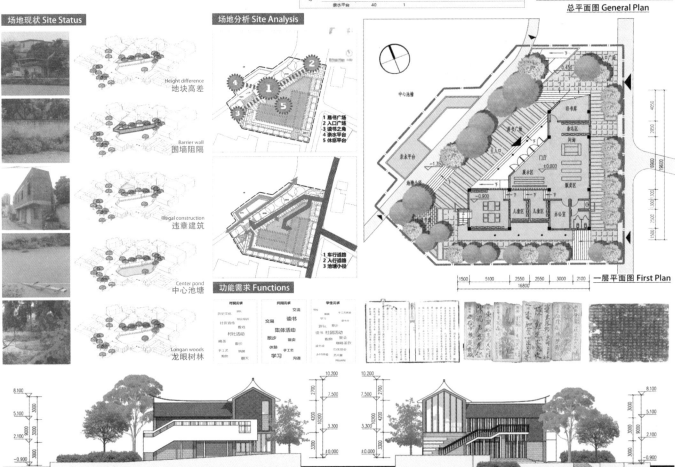

场地分析 Site Analysis

一层平面图 First Plan

南侧立面图 South Side Elevation

北侧立面图 North Side Elevation

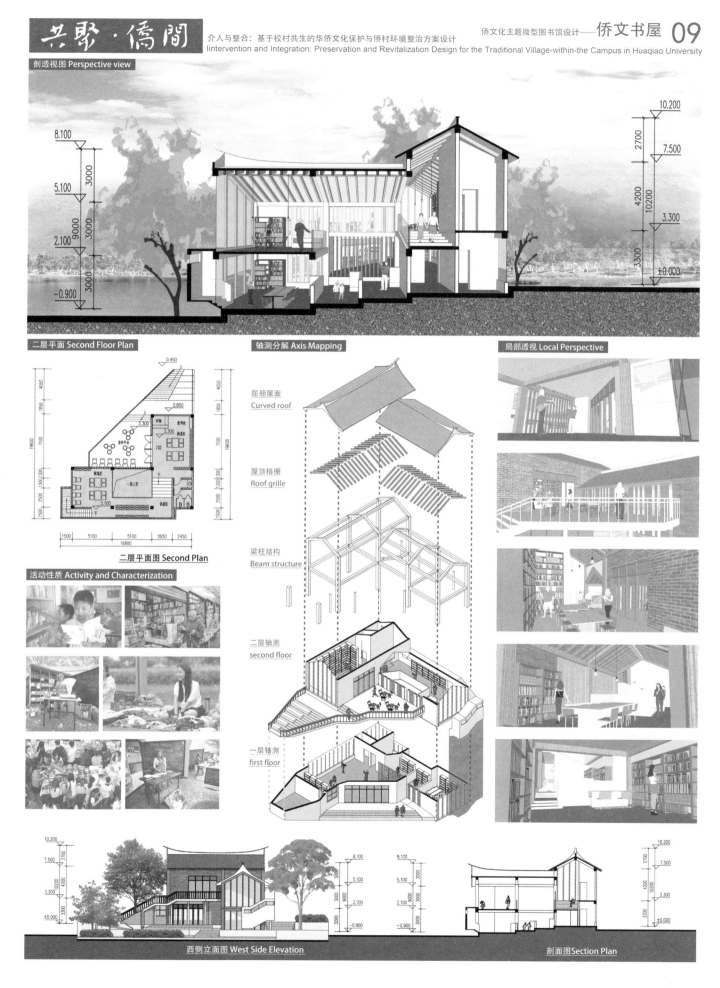

融合·共生与介入·整合—— 华侨大学厦门校区与村落的城市设计研究

織補僑垣

補垣织侨 PATCH THE TEXTURE & WEAVE THE CULTURE

指导老师：邓蜀阳 阎波
设计者：曹雅涵 何格 张宇

教师评语：
作品发现了场地风貌较保护不力、建筑风貌缺失、新建建筑文化建筑风格缺微等问题，采取针灸设计介入方法，选取典型村落通过道路梳理、修复并植入文化活动等方式进行肌理修补，制定风貌整治指引、文化活动指引，激活侨的空间与文化环境，以期校村内差异、新建传统侨文化不均等问题，采取针灸设计方法，选取典型村落保护补貌的出文化活动承载面，最终实现校村共赢。

OVERALL STRATEGY ANALYSIS

EXISTING TRADITIONAL BUILDINGS AND WALLS

PUBLIC SPACES IN VILLAGE
- 西珩村中心
- 南尾井村中心
- 李氏宗祠
- 潘涂村中心
- 宅内村中心
- 下蔡村中心

SUPERPOSITION OF THE ROUTES

PUBLIC SPACES IN CAMPUS
- 操场
- 休闲健身区
- 室外篮球场 室外羽毛球
- 室外篮球场
- 龙舟赛广场

SUPERPOSITION OF PUBLIC SPACES

POINTS WHICH CAN BE ACTIVATED

DISTRICT STRATEGY ANALYSIS

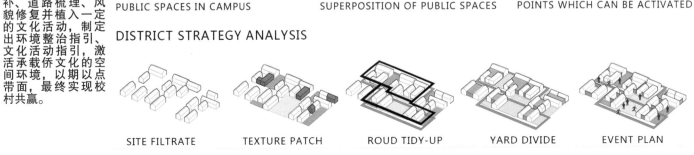

SITE FILTRATE | TEXTURE PATCH | ROUD TIDY-UP | YARD DIVIDE | EVENT PLAN

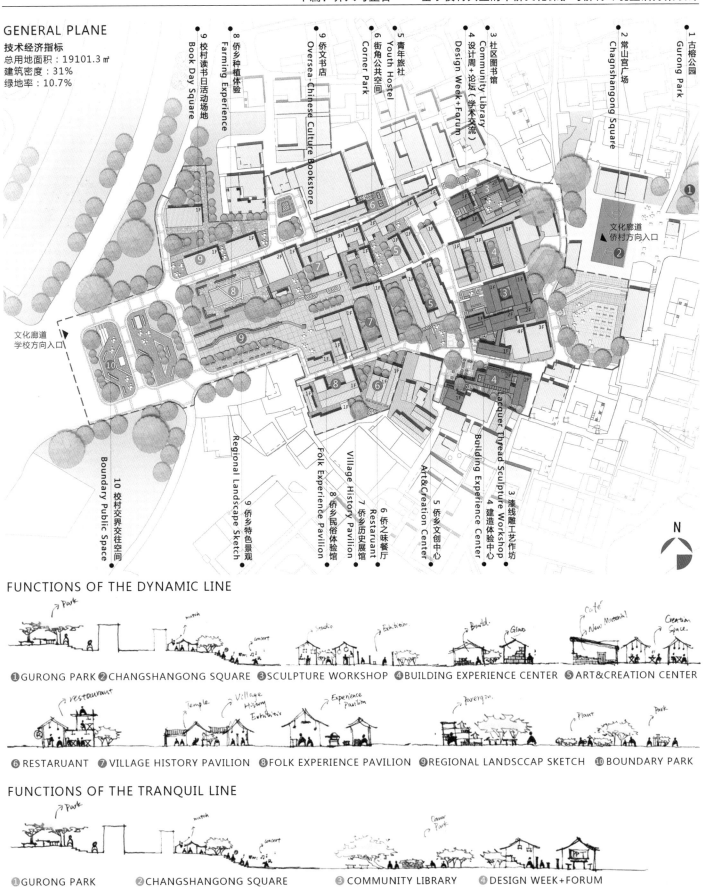

结构分析 STRUCTURE ANALYSIS

路网结构 ROAD NETWORK STRUCTURE

功能分区 FUNCTIONAL PARTITIONING

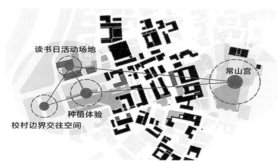
关键节点 IMPORTANT POINTS

单体建筑 IMPORTANT BUILDINGS

要素对比 ELEMENT COMPARISION

场地肌理 Site Texture

Current

用地面积 19101.3 ㎡

After Design

用地面积 19101.3 ㎡

建筑肌理 Building Texture

Current

待拆建筑 1320 ㎡
保留建筑 4164.2 ㎡

After Design

总建筑面积 6705.6 ㎡

院落分布 Yard Distribution

Current

面积 717.0 ㎡　　占地率 3.7%

After Design

面积 2252.6 ㎡　　占地率 11.8%

绿地分布 Vegetation

Current

面积 1725.5 ㎡　　占地率 9.0%

After Design

面积 2045.8 ㎡　　占地率 10.7%

广场分布 Square Distribution

Current

面积 280.0 ㎡　　占地率 1.4%

After Design
面积 3633.2 ㎡　　占地率 19.0%

道路 Roud Network

Current

面积 1230.3 ㎡　　占地率 6.4%

After Design
面积 5132.2 ㎡　　占地率 26.8%

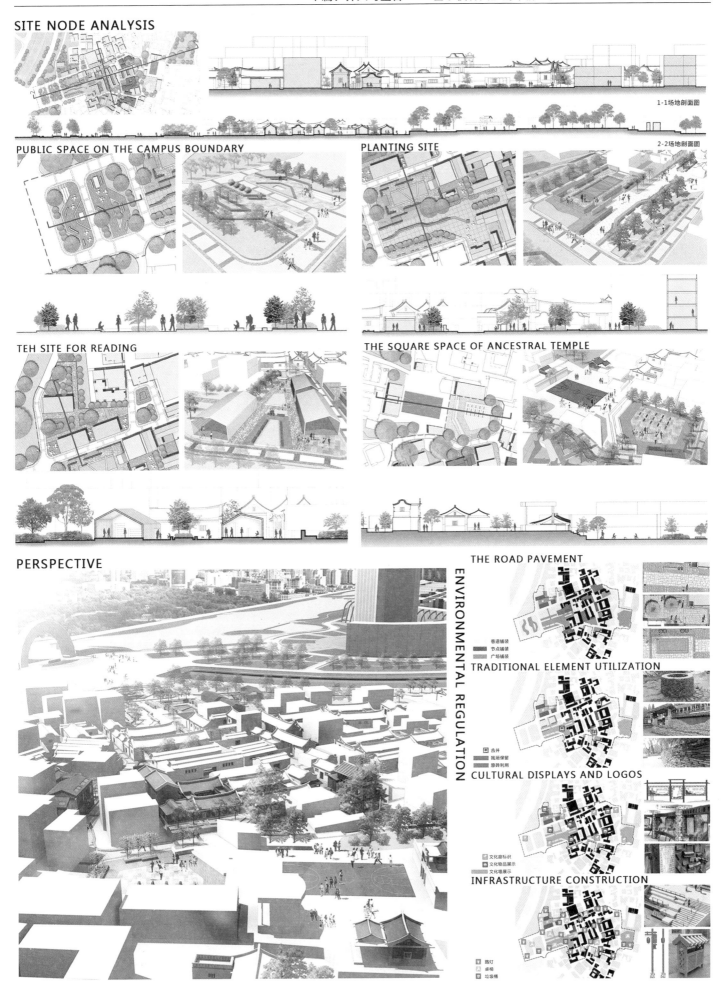

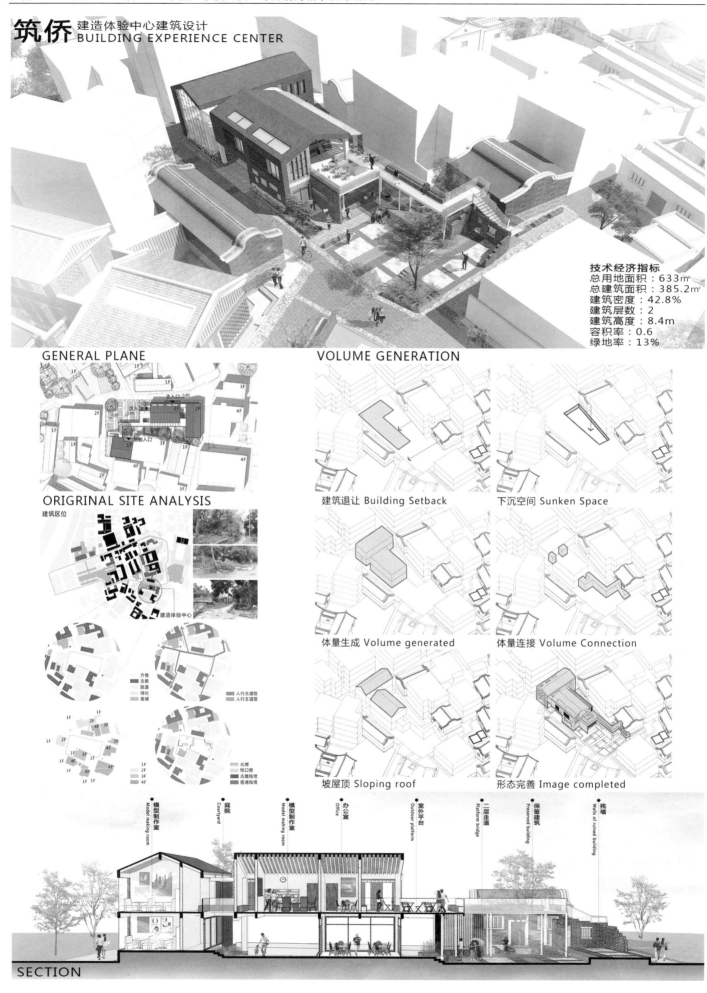

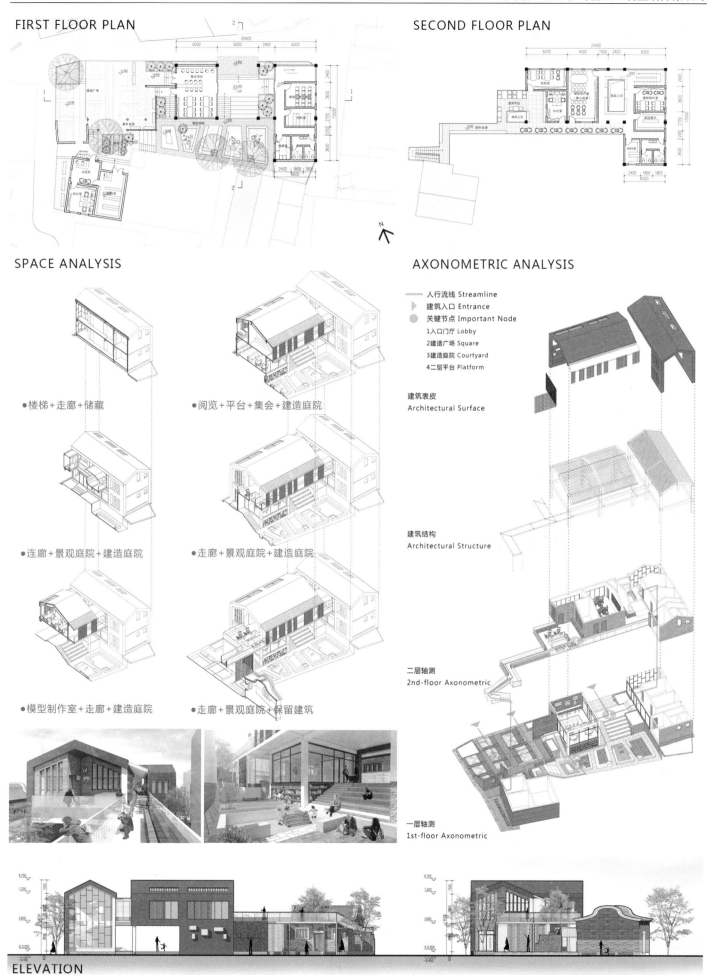

融合·共生与介入·整合——华侨大学厦门校区与村落的城市设计研究

侨脉相承 LACQUER THREAD SCULPTURE CRAFT WORKSHOP
漆线雕工艺作坊

概念构思 CONCEPT

梳理脉络 延续肌理 — 脉
传统工艺 传承文化 — 侨

侨脉相承

工艺简介 CRAFT INTRODUCTION

漆线雕
漆线雕发源于福建泉州，是中国漆艺文化宝库中的艺术瑰宝之一，是闽南地区独有的传统工艺。

制作流程 PROCESS

材料制作 Material Making　设计底稿 Design Draft

搓线 Rolling Line → 盘绕形体 Coiling Line

贴金箔 Gilding ← 上明漆、粉白土 Japanning

形体生成 FORM GENERATION

建筑体量连接 Building body mass connection　竖向高差关系处理 Vertical difference relation　坡屋顶建筑造型 Slope roof and grating

总图分析 PLAN ANALYSIS

新旧建筑 Building　内外院落 Courtyard　轴线关系 Axis

■新建筑　■老建筑　■联接体
入口广场　内院　下沉庭院

新建筑联系传统建筑 New buildings link traditional buildings　建筑围合出不同的院落 Different courtyards enclosing buildings　建筑之间存在轴线关系 There are many axis between buildings

总平面图 PLAN

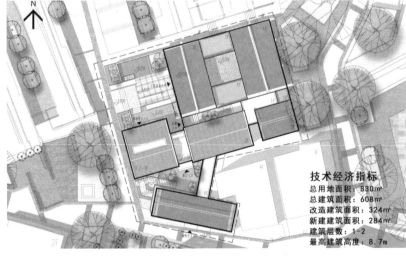

技术经济指标
总用地面积：830㎡
总建筑面积：608㎡
改造建筑面积：324㎡
新建筑面积：284㎡
建筑层数：1-2
最高建筑高度：8.7m

立面图 ELEVATION

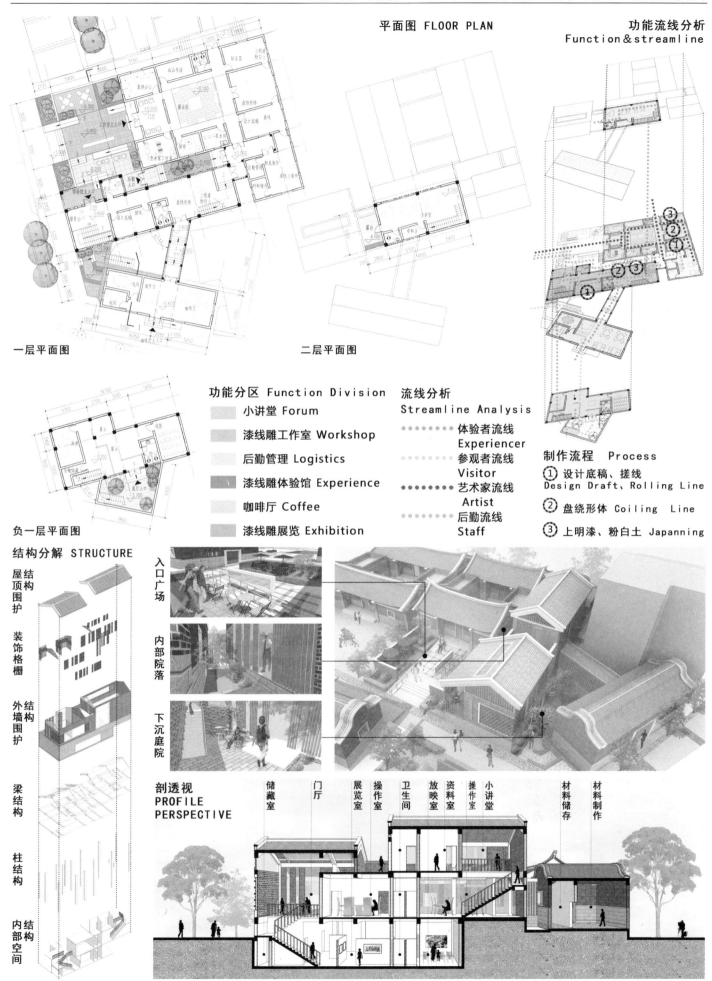

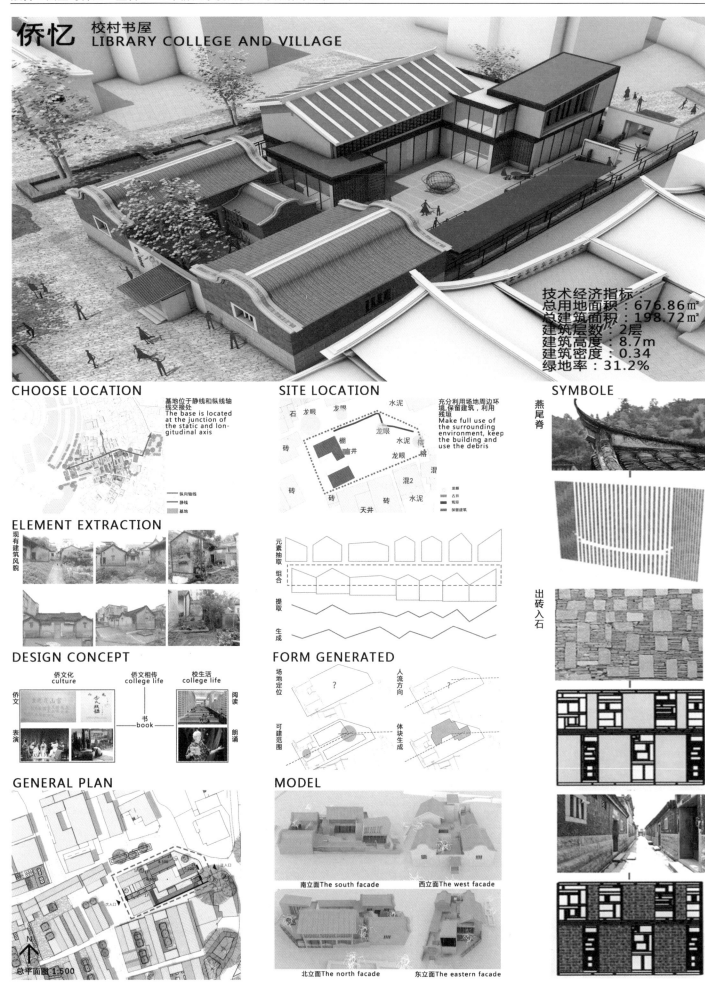

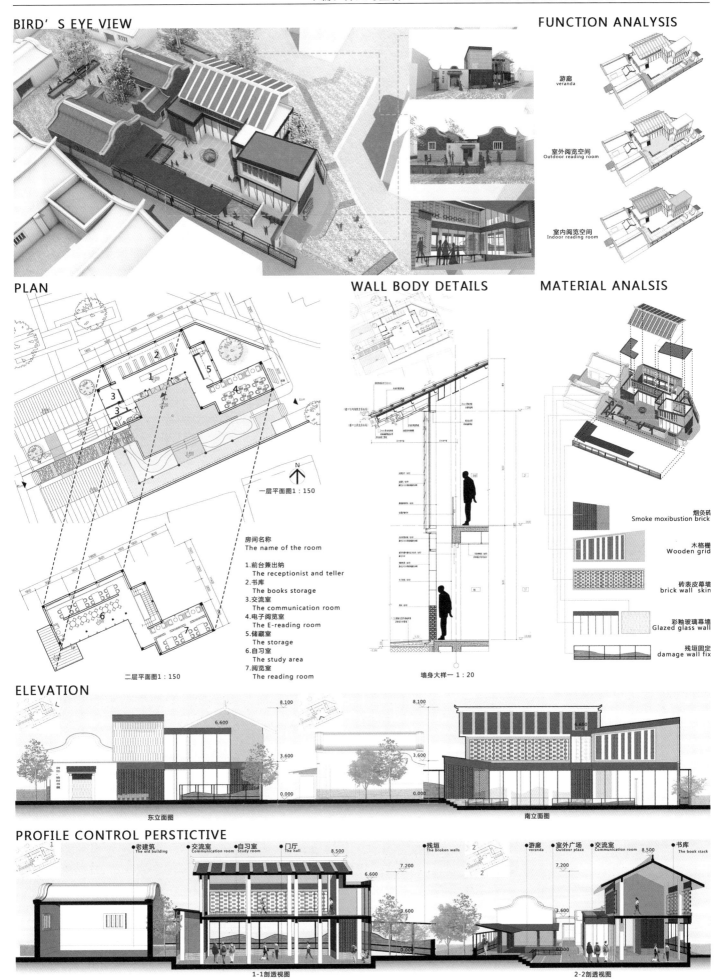

融合·共生与介入·整合——华侨大学厦门校区与村落的城市设计研究

戏致入微
INTERVENTION AS IMMERSION

侨村中心整治设计 OVERSEAS CHINESE VILLAGE CENTER REGULATION DESIGN
介入与整合——基于校村共生的侨村文化保护与侨村环境整治方案设计
Intervention and Integration: Preservation and Revitalization Design for the Traditional Village-within-the Campus in HQU

总体设计 01

指导老师：邓蜀阳　阎波
设计者：张叶青　汪天琦　邱鞠

前期分析 BACKGROUND RESEARCH

概念生成 CONCEPT ANALYSIS

功能介入 Function intervene	文化介入 Culture intervene	活力介入 Vitality intervene	整合 Integration
校村纽带 Contact university and village	游神路线 Pageant on immortals	村中心 Village center	中心辐射 Village center radiation

改造策略 RENEWAL STRATEGY

BEFORE　道路 Road　建筑 Buildings　开放空间 Open Space　景观 Landscape
AFTER　道路 Road　建筑 Buildings　开放空间 Open Space　景观 Landscape

活动植入 ACTIVITIES STRATEGY

教师评语：
作品紧紧抓住非物质文化要素与物质空间、景观的关系，把传统戏曲活动、侨文化中的祭祀与宗教活动的载体空间以及共生的文化空间、公共空间、村落环境结合起来，提出了具体的整治方案和活动指引，以最大限度地保护侨文化生存的土壤。作品侨演、祭祀生活空间、建筑空间、民俗表演空间、街巷空间等物质文化与非物质文化之统一表现。

198

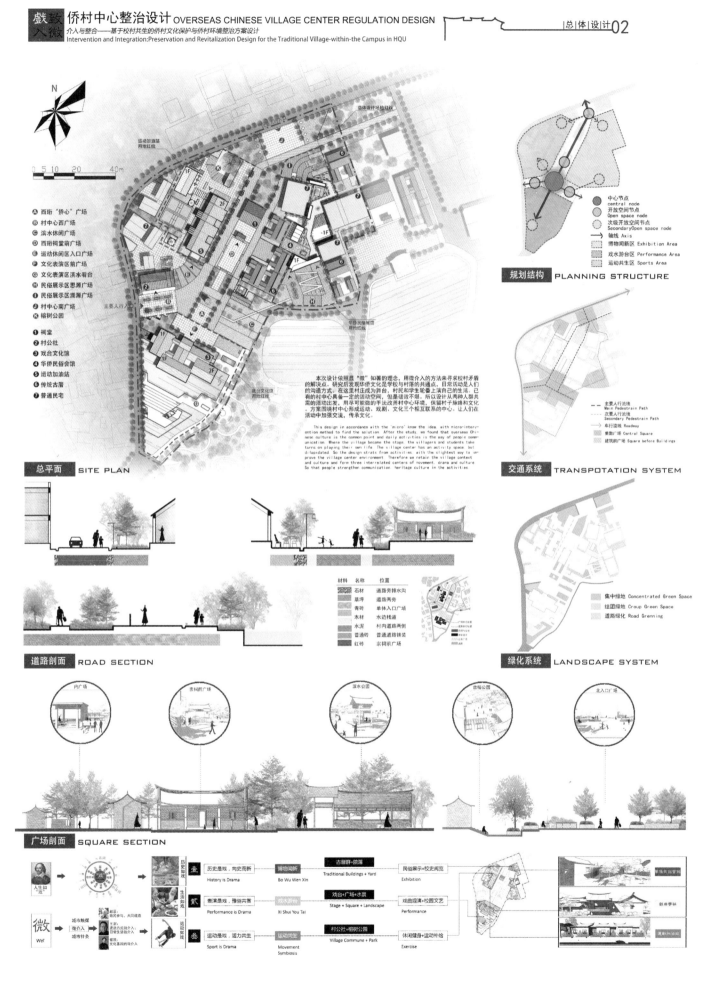

融合·共生与介入·整合 —— 华侨大学厦门校区与村落的城市设计研究

戏致入微 华侨民俗会馆建筑方案设计 HUAQIAO FOLK MUSEUM DESIGN
介入与整合——基于校村共生的侨村文化保护与侨村环境整治方案设计
Intervention and Integration: Preservation and Revitalization Design for the Traditional Village-within-the Campus in HQU

博|物|闻|新 03

入口效果图 RENDERING OF ENTRANCE

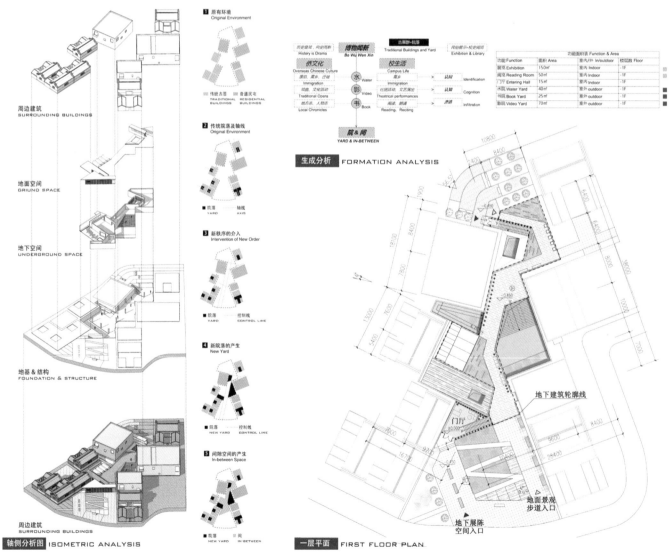

轴侧分析图 ISOMETRIC ANALYSIS　　一层平面 FIRST FLOOR PLAN

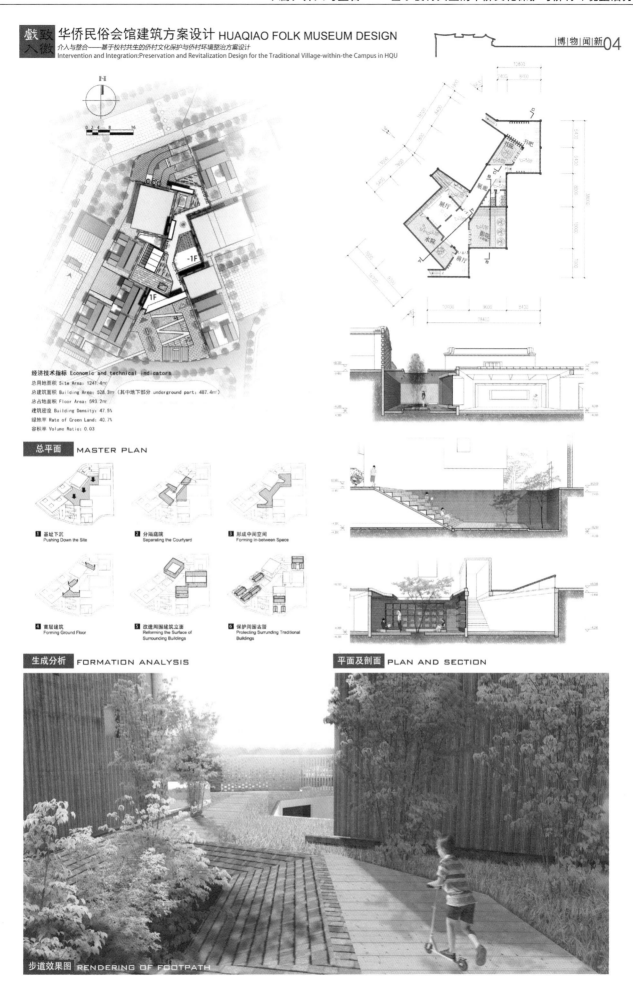

华侨民俗会馆建筑方案设计 HUAQIAO FOLK MUSEUM DESIGN
介入与整合——基于校村共生的侨村文化保护与侨村环境整治方案设计
Intervention and Integration: Preservation and Revitalization Design for the Traditional Village-within-the Campus in HQU

 |戏|水|游|台|05

位置 SITE

轴线 AXIS
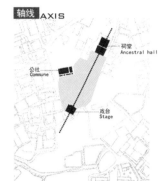

现状 SITUATION / 轴线剖面 THE PROFILE ALONG THE AXIS

总平设计 SITE DESIGN
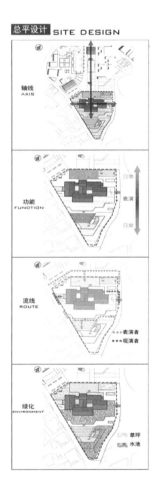

总平面图 GENERAL LAYOUT
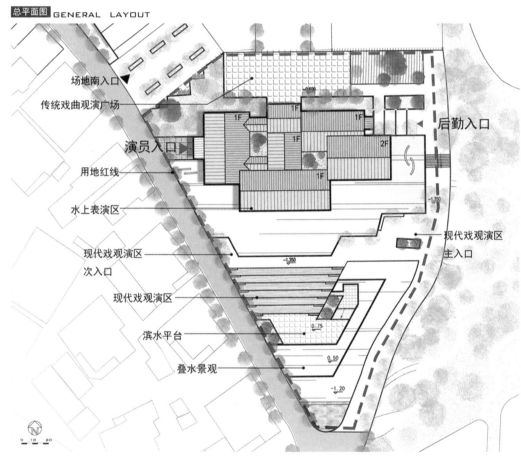

戏台设计 STAGE DESIGN

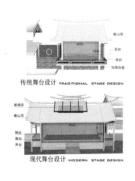
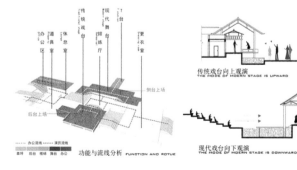

下篇：介入与整合——基于校村共生的华侨文化保护与侨村环境整治方案设计

戏曲学社 THE DRAMA SOCIETY

介入与整合——基于校村共生的侨村文化保护与侨村环境整治方案设计
Intervention and Integration: Preservation and Revitalization Design for the Traditional Village-within-the Campus in HQU

|戏|水|游|台|06

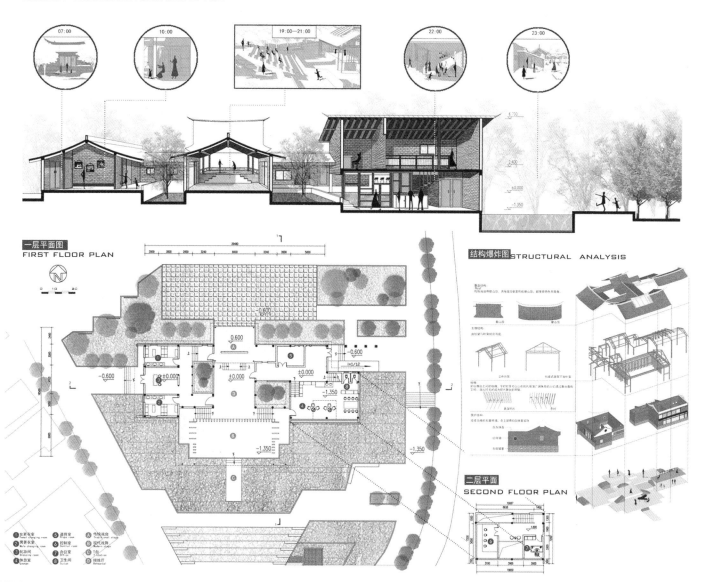

1-1 剖透视 1-1 SECTONAL PERSPECTIVE VIEW

一层平面图 FIRST FLOOR PLAN

结构爆炸图 STRUCTURAL ANALYSIS

二层平面 SECOND FLOOR PLAN

透视图 PERSPECTIVE

设计说明 DESIGN NOTES

本方案是西村中心戏台的原址重建项目。所谓"戏如人生"，戏台获得村校的共鸣。本设计始于"戏"，寻求"戏"与人、空间的关系。村戏是传统戏曲的代表，校戏是现代表演的代表。根据二者不同的表演需求，用两个相背而坐的舞台设计不同的表演形式，结合地形形成两种观影方式，让舞台具有承载多种表演的可能。校方和村民得到更多交流的机会。在这里，不再有校村之隔，只有观众和演员。

The program is the reconstruction of the center of the West Village. As the saying goes "play like life", play can get the resonance of the village and school. The design began in the "play", seeking the relationship between "play" and people or space. Village opera on behalf of the traditional opera, school drama on behalf of modern performances. According to its different requirement in stage, play acting and viewing, the plan uses two stage to provide different performances, combined with terrain to create two kinds of viewing methods. So that the stage has the possibility of carrying a variety of performances, and providing more opportunities for communication. There is no more school and village, only the audience and the actors.

功能分区 Function	表演区域	面积/Area	数量/Quantity
传统戏曲舞台 Traditional stage	表演区 Performance area	40*60㎡	1
	候场区 Waiting room	30㎡	1
现代戏剧 Modern stage	表演区 Performance area	50*30㎡	1
	候场区 Waiting room	25㎡	1
	控制室 Control room	32㎡	1
后台 （两个舞台共用） Backstage (Two-stage sharing)	化妆间 Dressing room	17㎡	1
	更衣室 Locker rooms	15㎡	2
	卫生间 Toilet	10*10㎡	2
	休息厅 Rest hall	30㎡	1
	办公室 Office	12㎡	1
	道具间 Props room		1

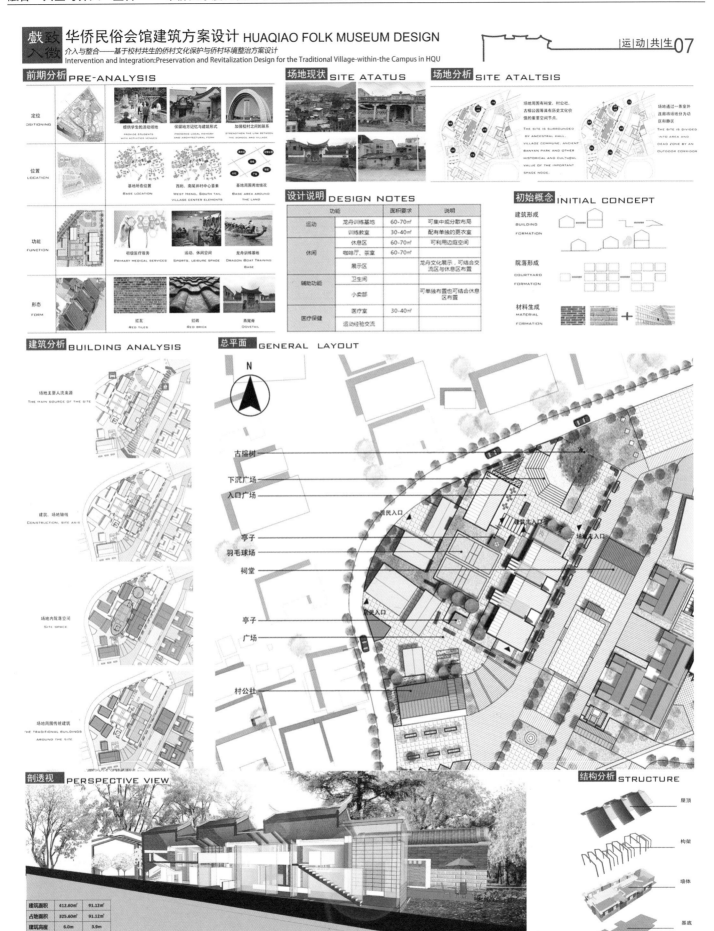

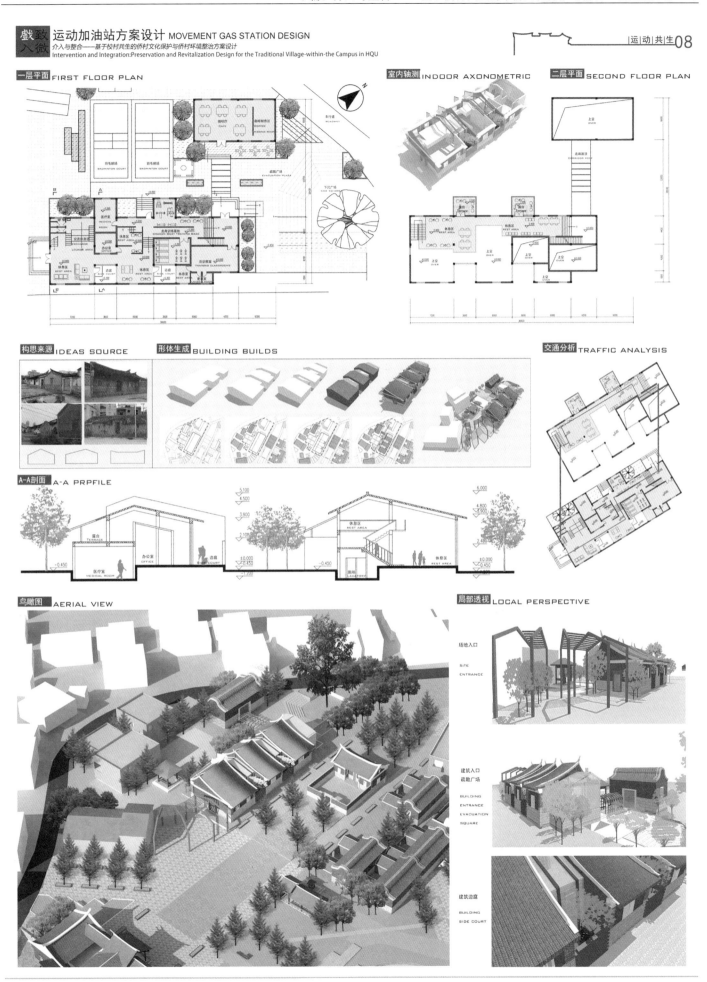

下篇：介入与整合 —— 基于校村共生的华侨文化保护与侨村环境整治方案设计

学生团队 Student Team

靳柳

缅金娣

齐大禹

王欣

马步青

沈安杨

龚忱

罗文婧

蕾朗珠

融合·共生与介入·整合——华侨大学厦门校区与村落的城市设计研究

二次生长
SECONDARY GROWTH

指导老师：张路峰　课程助教：王大伟　程力真
设计者：靳柳　杨金娣　齐大勇

调研分析

村落生长载体
Village Growth Point

村落的生长载体有很多，经调研和分析我们提取出与村落生长关系最紧密的四个载体——祠堂/古厝/宫庙/榕树

There are many growth carriers in the village. After research and analysis, we have extracted four carriers that are closely related to the growth of the village – the ancestral hall / the ancient house / the temple / the banyan tree

兑山村族谱 关系村子重要的社会结构，李氏家庙为全村的祖祠，由祖祠分出各社的祠

全村信仰保生大帝、关帝；每个自然村还有属于各自的守护神，并建有神庙

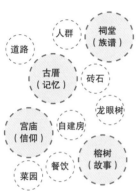

村落重要生长点格局分析
Growth point analysis

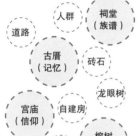

有谱
Family Tree Pedigree

宗祠与空地，或宗祠、空地与戏台组成

有信仰
Belief

宫庙、香炉，或宗庙、香炉与戏台组成

华侨和村民祭祖活动、婚庆活动
Ancestor Worship

祭神、游神活动
Worship

教师评语：
设计旨在通过"微营建"精准开启村落更新与转型的进程，立意鲜明，有较强的操作性。设计选择具有"公地"特征的遗弃老屋、路侧荒地，分别改造为校村共享的文化活动中心、移动式集装箱营地、文化展览长廊，各设计单元都实现了多功能复合与弹性使用的目标。设计将校园功能植入村落，并以共享的方式推动村民与师生交流合作。建筑形式考虑了村落的空间尺度和色彩特征，环境营建与村落交通和景观融为一体，基本达到了设计要求。

祠堂分布
祠堂分布在各社，多处在社中心

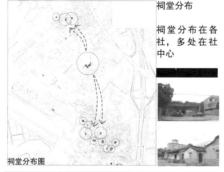

祠堂分布图

宫庙分布
最大的宫庙为金鞍山寺，在回迁房区其余分布在各社

现状照片

宫庙等级图

总结

有谱 – 生长的起点

有谱的载体是祠堂，祠堂的背后反映着村子的社会关系，社会关系是村子生长的起点

当前问题

祠堂在集会的时候才会有使用，平常很少人使用，空间缺少活力，反映出村民交流很少

总结

有信仰 – 生长的支撑

村落生长的核心是人，兑山村人们的信仰是人们的精神支撑

当前问题

游神路线参与度少，背后反映出村民的文化生活，参与度的减少反映出村子凝聚力的减弱

下篇：介入与整合——基于校村共生的华侨文化保护与侨村环境整治方案设计

是闽南最有代表意义的传统建筑，承担着村民的居住功能

榕树被评为福建省省树，具有普遍性、认同感，以及一种独特的城市归属感和自豪感

华侨大学于1960年由国家创办，是一所为方便华侨青年回国，升学而设的大学与华侨文化密不可分

有记忆
Memory

单层，双坡屋顶，屋顶有燕尾脊和马鞍脊两种形式

居住活动
Living

古厝分布
古厝分布在村子内围，部分保存良好，多数不再使用

现状照片

古厝分布图

有故事
Story

树冠巨大，直径 6~20m 不等

娱乐活动
Entertainment

榕树分布
多数分布在村落中，少数分布在已建校园内

现状照片

榕树分布图

华侨大学
Huaqiao University

体量较大，与村落体量差别大

学生活动
Student Activity

校区尺度
校区内多为大体量建筑，道路较宽，整个学校尺度较大

现状照片

已建校区图底关系

总结

有记忆 – 生长的标志

村落主要功能为居住，承担居住功能的古厝数量反映着村落的生长

当前问题

古厝普遍分布在村内靠近中心的位置，集中分布。大部分古厝都存在不同程度的破损，且已经不再使用

总结

有故事 – 生长的见证

榕树作为当地存活时间最长的物种见证着村落的生长

当前问题

榕树下或者被绿地隔离，或者周围垃圾堆放，环境差

总结

华侨大学

华侨大学拥有丰富的资源，为村子质的更新提供契机

当前问题

华侨大学有一定的功能需求，需要进入村子，但当下对村子的影响是消极的，丰富的资源优势在村校融合上没有体现

总体概念设计 /General Design

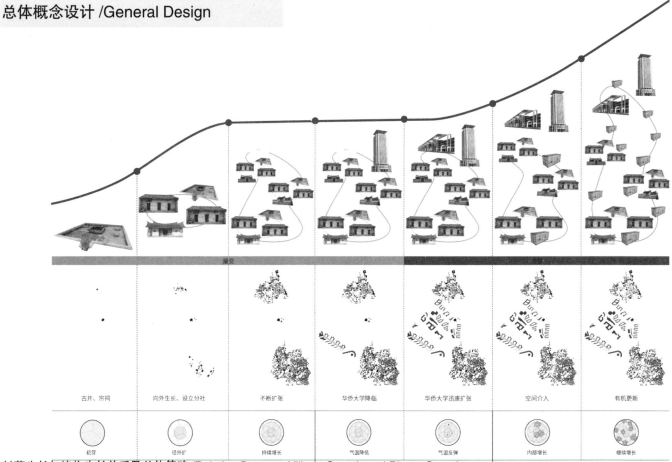

村落生长与植物生长关系及总体策略 /Relation Between Village Growth and Plants Growth

村子的前期是以量的增加为生长标志的过程，与植物生长有相似之处，我们借鉴自然界二次生长的概念对村落进行介入，通过对散布的生长点的整理，实现村子由量的增加转变为质的提高，从而实现村落的自我更新，让村子成为校园有机的部分，真正实现校村融合。

The early stage of the village is the process of increasing the number of signs of growth, and plant growth are similar, we learn from the concept of secondary growth of nature to intervene in the village, through the spread of our distribution points to achieve the village by the amount of the increase into a qualitative improvement, in order to achieve the village's self-renewal.

管理策略
Management Strategy

校方　第三方　村民
实施期 /Implementation Period

管理公司　校村
稳定期 /Stable Period

实施过程
Progress

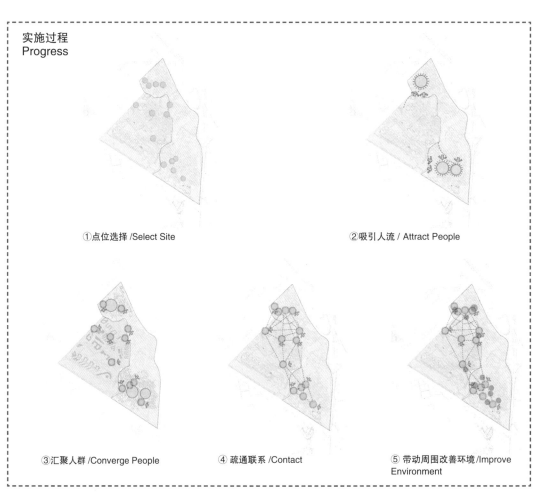

① 点位选择 /Select Site　　② 吸引人流 / Attract People

③ 汇聚人群 /Converge People　　④ 疏通联系 /Contact　　⑤ 带动周围改善环境 /Improve Environment

下篇：介入与整合——基于校村共生的华侨文化保护与侨村环境整治方案设计

介入手法 /Interventin Method

设计通过永久介入和临时介入两种手法对选取的空间节点进行处理。
永久介入：主要针对村中心进行永久性介入，强化其中心的空间存在感。
临时介入：主要针对村中的破败空间、无人使用空间、垃圾堆放点等消极空间。空间的处理主要目的在于通过集装箱的有效设置。

The design handles the selected spatial nodes by means of permanent intervention and temporary intervention.
Permanent intervention: mainly for the village center for permanent intervention, strengthen the center of the existence of space.
Temporary intervention: mainly for the village of dilapidated space, no use of space, garbage dump points and other negative space. The main purpose of space processing is through the efficient setting of the container.

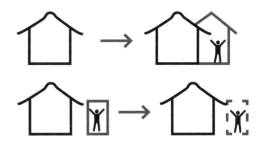

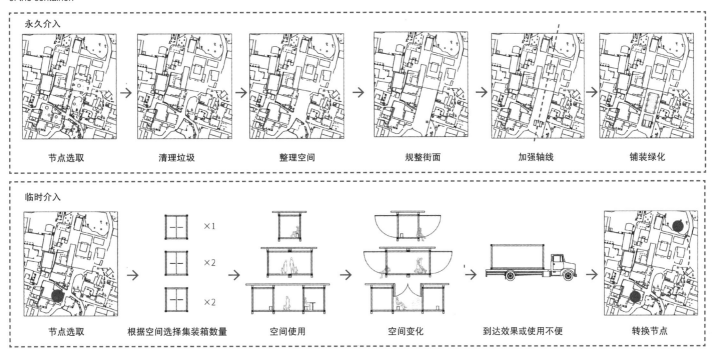

功能需求
Function Requirements

介入后总体策划图
Overall Plan

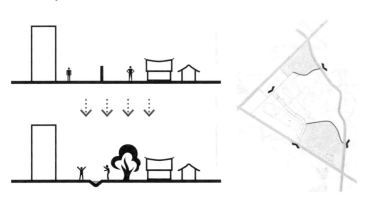

边界
Boundary

211

融合·共生与介入·整合 —— 华侨大学厦门校区与村落的城市设计研究

校·榕树公园新建方案设计

单体设计

项目位置 /Location

本设计旨在通过对场地的规划整理明确和加强西珩村中心的定位，同时将带有校园文化的社团活动中心引入其中，实现该区域品质和文化的提升，以此节点为重要生长点实现西珩社的二次生长的生长点。

The design on the one hand through the function into the activation of banyan near the space, into the school history as the campus carrying the culture of overseas Chinese culture, reflecting the overseas Chinese culture; the other hand, the main road side of the small volume building is conducive to find a sense of scale on campus.

总平面图 /Plan

透视图 /Perspective

东立面图 /East Elevation

一层平面图 /1st Floor Plan

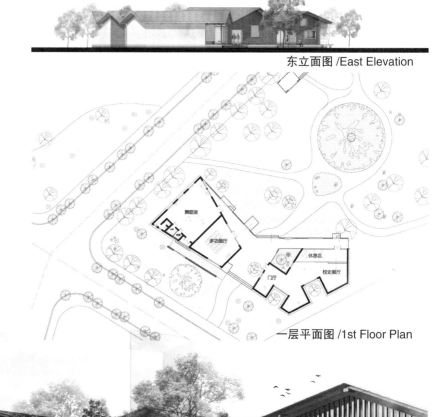

下篇：介入与整合——基于校村共生的华侨文化保护与侨村环境整治方案设计

单体设计

项目位置 /Location

改造前场地轴线关系图

改造前场地流线图

场地介入前轴测图

场地介入后效果图

透视图 /Perspective

一层平面图 /1st Floor Plan

通过介入改造提升该节点的空间品质，并疏通与南尾井社中心的通道，进而提升整个北区的品质和活力。

爆炸图 /Explosive View

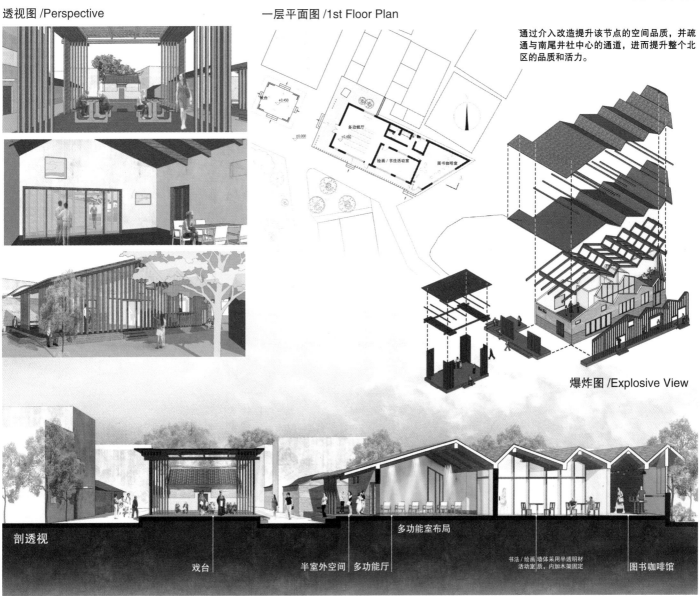

213

单体设计

鸟瞰图 Bird View of the Center of Zhainei

休闲书屋	儿童书店	戏台
集装箱组合方式	集装箱组合方式	集装箱组合方式
首层平面图	首层平面图	首层平面图
二层平面图	二层平面图	与祠堂轴线对应
节点	节点	节点

单体设计 Individual Design

下篇：介入与整合——基于校村共生的华侨文化保护与侨村环境整治方案设计

单体设计

常山宫节点鸟瞰图　Bird View of the Center of Pantu

| 设计工作室 | 集装箱组合方式 | 平面图 | 入口透视 |
| 咖啡吧 | 集装箱组合方式 | 首层平面图 | 二层平面图 |

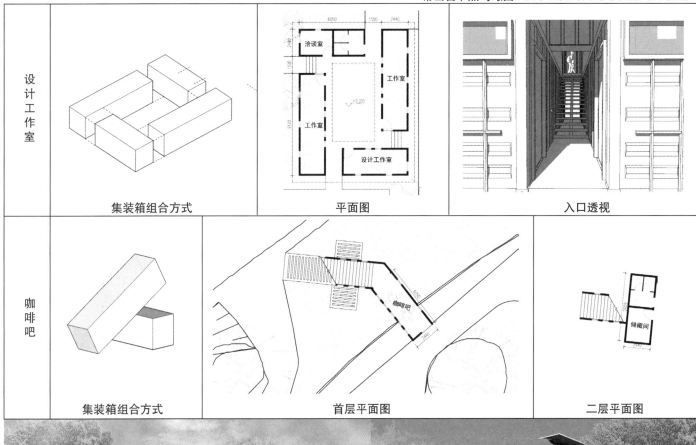

咖啡吧透视图　　　　　　　　　咖啡吧入口透视图

215

融合·共生与介入·整合 —— 华侨大学厦门校区与村落的城市设计研究

单体设计

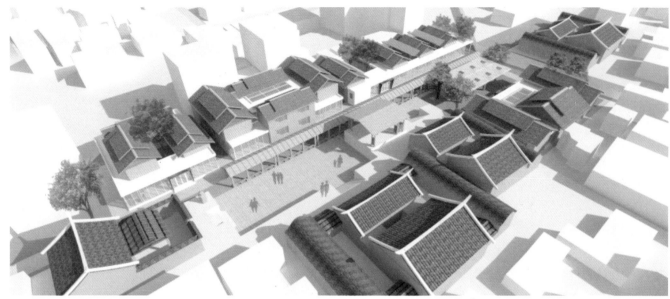

鸟瞰图　Bird View of the Center of Zhainei

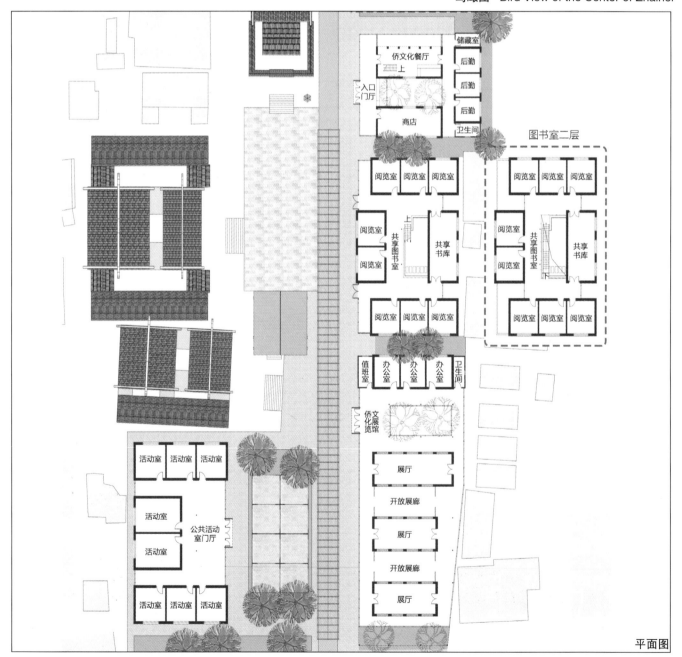

平面图

下篇：介入与整合——基于校村共生的华侨文化保护与侨村环境整治方案设计

单体设计

鸟瞰图　Bird View of the Center of Zhainei

侨文化展览馆剖透视图

茶餐厅剖透视图　　　　　　　　　　茶餐厅立面图

217

融合·共生与介入·整合 —— 华侨大学厦门校区与村落的城市设计研究

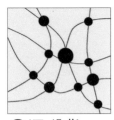

通脉
CONNECT THE VEIN

指导老师：张路峰　课程助教：王大伟
设计者：王欣　马步青　沈安杨　程力真

建筑发展现状
Present Situation

■ 祠堂和古厝
■ 可利用
■ 可新建
■ 其他

校村现状肌理

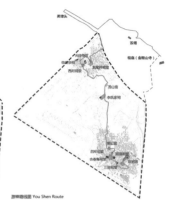

道路现状
Road Situation

停车现状
Parking Situation

学校道路　村落车行道　村落步行道

广场公园　　私人角落　　路边空地
分布在祠堂、古榕公园等开阔地，侵占了广场等公共空间。　分布住户院落或附近小型平坦场地。　分布在较宽机动车道旁的空地。

村落发展过程
Development Process

教师评语：
　　本组设计认定交通一体化是校村融合的起点，村落交通结构的破碎和景观结构的障碍。设计的交通节点建筑能成为校村融合的共生。设计的交通络体系包括选线，步行与非机动车交通混行路面，重要的交通节点建筑改造。选定线路将校园周边村落的祠堂、寺庙、戏台、古树、广场、古民居群、池塘连为一体，巧妙形成村落文化景观展览路线。节点建筑包括文化博物馆、活动中心等建筑群，都是通过老屋改造实现，也基本实现了节点建筑的设计目标。

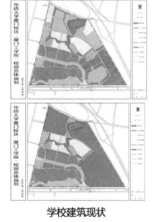

学校建筑现状

校村争地

缺少
运动场地
教学用地
公共设施用地
学生居住用地
周转居住用地

村中心现状
Village Center Situation

华侨特色文化
Hua Qiao Culture

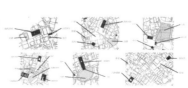
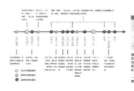

下篇：介入与整合——基于校村共生的华侨文化保护与侨村环境整治方案设计

概念生成
Concept Formation

人体 ➡ 村落
脉络 ➡ 华侨文化
穴位 ➡ 村中心

兑山村的华侨文化犹如它的脉络，脉络不通，则村子不活。疏通文化脉络的关键在于找准穴位，即村中心。

总体策略
Strategy

策略说明：
- 村中心的开敞空间多被停车占用。
- 疏通村落文化脉络的第一步在于疏解村中心的停车压力。
- 还原村中心的开敞空间后，运用中医针灸手法进行点式介入。

路网规划
Road Plan

校园路网现状　　村落路网现状　　路网规划图　　停车点规划图

项目介入设计
Project Design

游神路线铺地设计
Pavage Design

教学区：综合教学楼、实验楼、图书馆
生活区：兰苑、梅苑、刺桐苑、凤凰苑、紫荆苑、紫荆餐厅、凤凰餐厅
文博区：华文学院、廖承志音乐厅、华侨博物馆、校史馆、档案馆
运动区：运动场、篮球场、体育馆

■ 已建成
■ 未建成（拟选）

单元式停车棚设计
Parking Shed Design

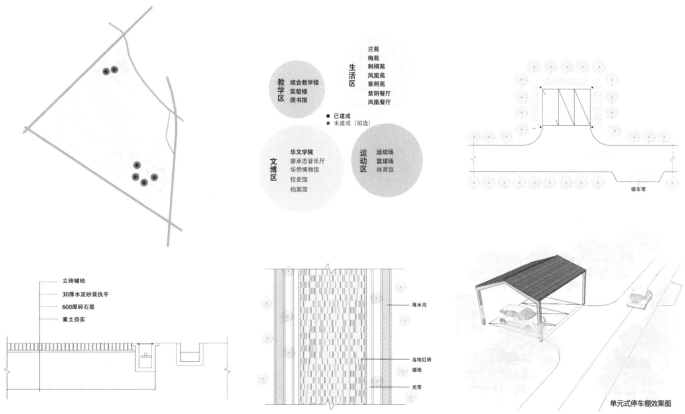

立砖铺地
30厚水泥砂浆找平
600厚碎石层
素土夯实

雨水沟
当地红砖
铺地
光带

单元式停车棚效果图

融合·共生与介入·整合 —— 华侨大学厦门校区与村落的城市设计研究

项目总平面图
Site Plan

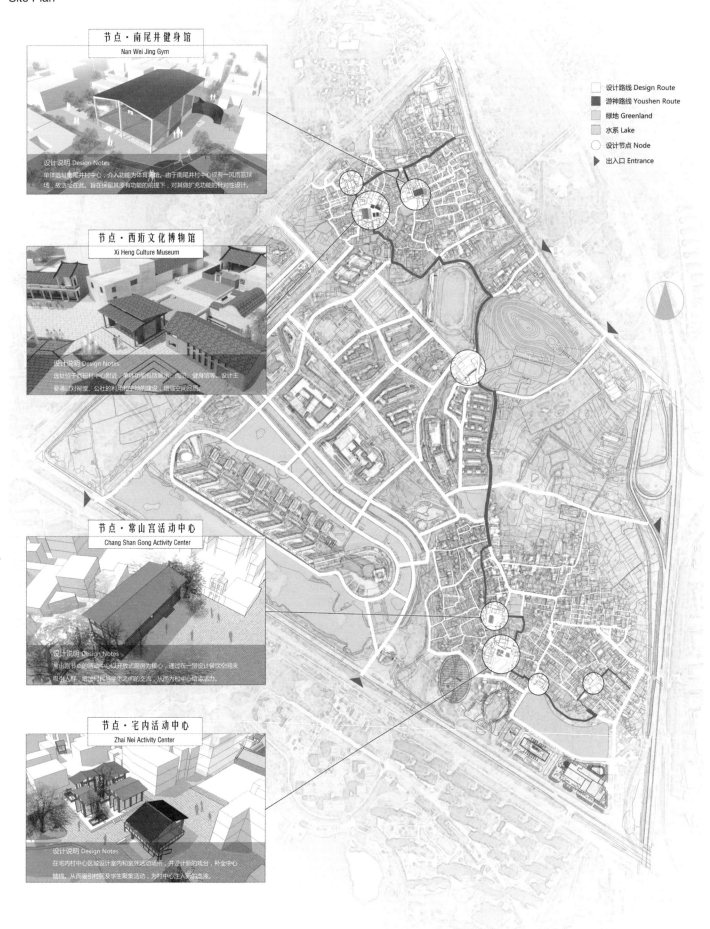

下篇：介入与整合——基于校村共生的华侨文化保护与侨村环境整治方案设计

西珩

选址位于西珩村祠堂附近，这里是西珩村的村中心，前有广场，后有戏台。
The design is located in the center of Xiheng village, near the ancestral temple. There is a square before it and a stage behind it.

总平面图
Site Plan

戏台 / 凉亭
Opera Stage

西立面图
Western Elevation

南立面图
South Elevation

戏台平面
Plan for Opera

休憩平面
Plan for Relax

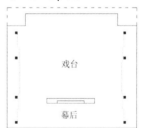
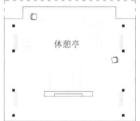

剖面图 1-1
Section 1-1

戏台构成
Composition

西珩戏台轴线剖透视
Perspective Section

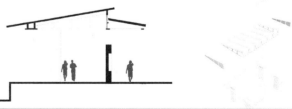
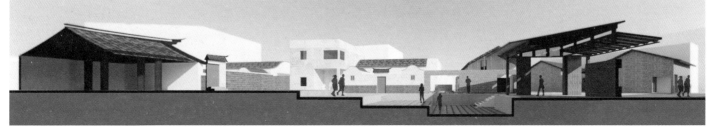

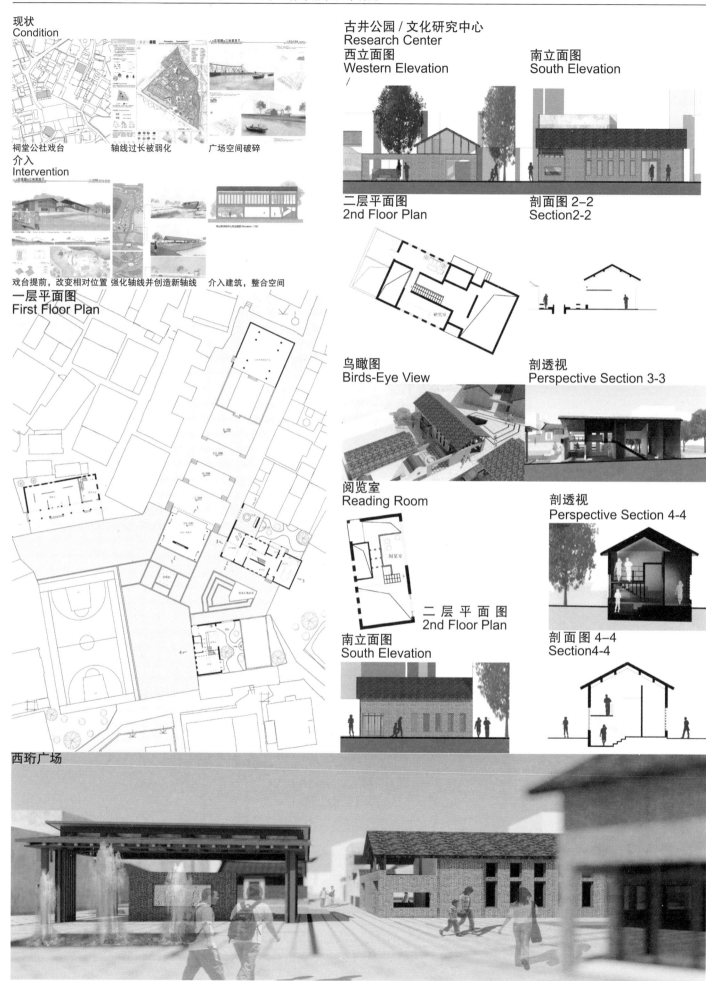

下篇：介入与整合 —— 基于校村共生的华侨文化保护与侨村环境整治方案设计

南尾井体育馆设计
Nanweijing Gym Design

南尾井

选址位于南尾井村中心附近的风雨篮球场。

现状照片 Situation Photos

总平面图
Site Plan

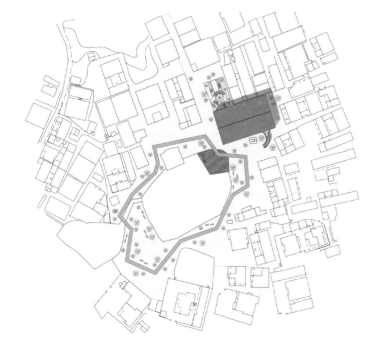

西立面图
Western Elevation

南立面图
South Elevation

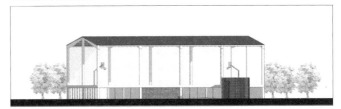

剖面图
Section

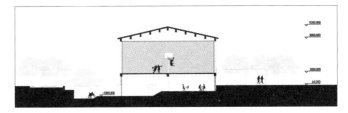

鸟瞰图
Aerial View

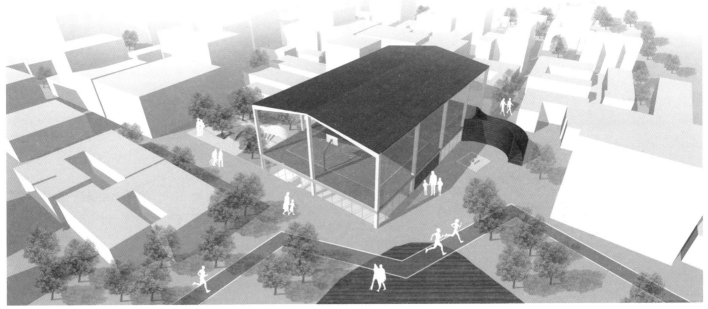

融合·共生与介入·整合 —— 华侨大学厦门校区与村落的城市设计研究

平面图
Plan

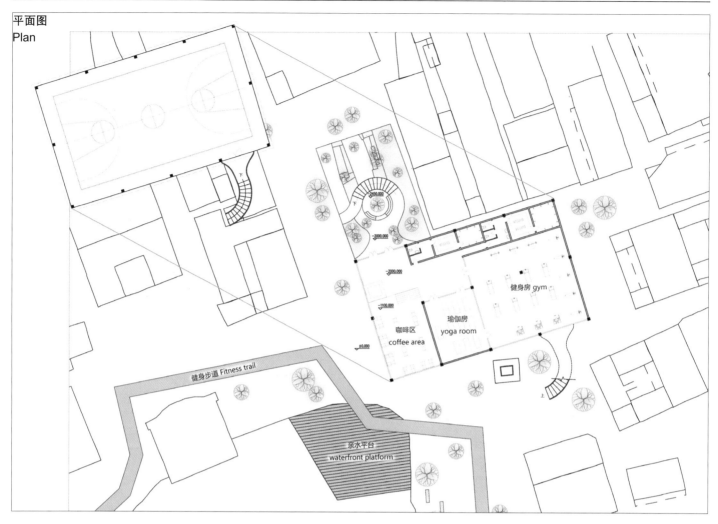

西珩体育馆设计
Xiheng Gym Design

平面图　　　　　　　　　　　　　　　　　　　总平面图
Plan　　　　　　　　　　　　　　　　　　　　Site Plan

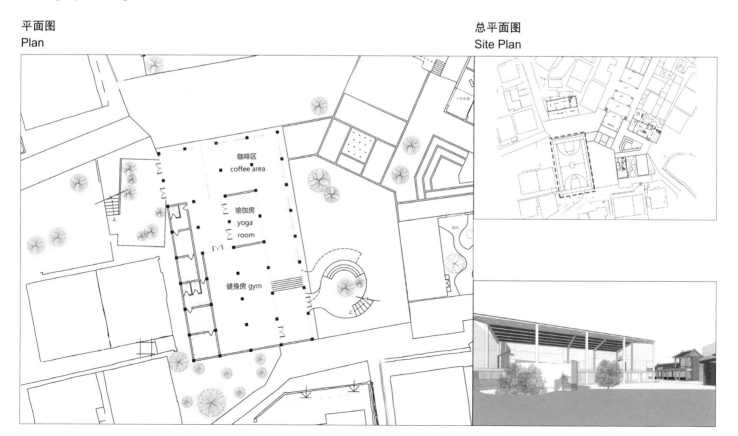

下篇：介入与整合 —— 基于校村共生的华侨文化保护与侨村环境整治方案设计

基地位置图
Location

总平面图
Site Plan

平面图
Plan

西立面图
Western Elevation

南立面图
South Elevation

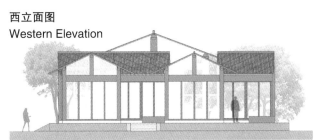

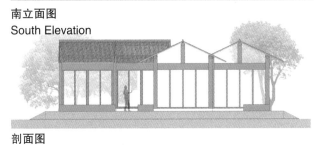

剖面图
Section

鸟瞰图
Aerial View

融合·共生与介入·整合 —— 华侨大学厦门校区与村落的城市设计研究

基地位置图
Location

总平面图
Site Plan

平面图
Plan

立面图
Elevation

剖面图
Section

功能分解图
Functional Diagram

鸟瞰图
Aerial View

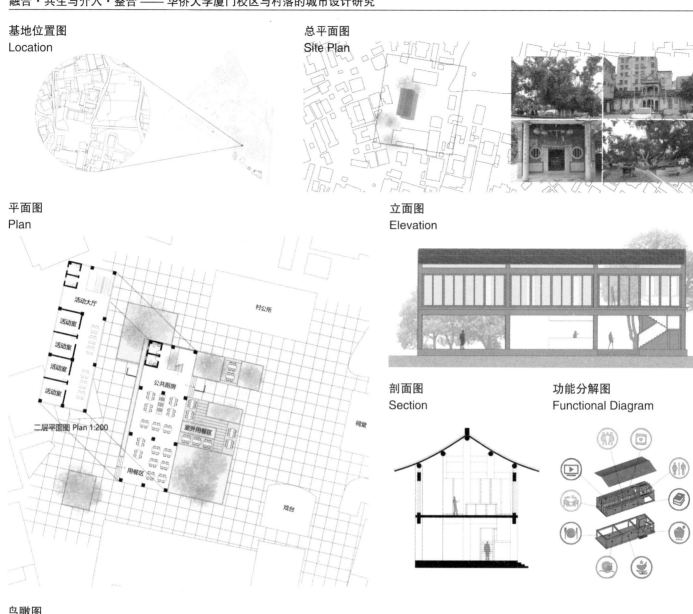

下篇：介入与整合 —— 基于校村共生的华侨文化保护与侨村环境整治方案设计

总体设计
Integrated Design

在华侨大学厦门新校区动工以前，校园的所在地是一片普通的闽南村落。随着这一片土地被确定为华侨大学校园建设用地，新的城市的体系闯入了原有的村落空间中。由于资金的限制，华侨大学无法一次征收所有的农村土地和房屋，因而首先学校选择了在没有村落的低洼地带建设校园建筑。自那以后，学校见缝插针进行了大量的建设。到今天为止，这个进程仍然在继续。因而华侨大学的用地范围内，存在着学校和村落两种聚落形式，它们之间的一条围墙将学校的土地分成了两部分。

The site is located in Xiamen, Fujian Province, and there was a village on the site. The municipal government gave the site to the Huaqiao University, since then, the university began to construct its campus. Due to financial problems, the university can not afford to demolish the village, so the construction started at where no houses are built. Gradually, the university occupied the spaces, within the limit of money. This process is still continuing today, and as a result, the site is decided by a wall into 2 parts: university campus and village.

■ 卫星影像 Satellite Image

■ 建筑 Buildings　■ 边界 Border

绿网　GREEN NETWORK

指导老师：张路峰
课程助教：王大伟
设计者：蔡忱　罗文婧　雷明珠　程力真

■ 传统建筑现状 Traditional Building Conditions

■ 地表水系 Surface Water

■ 可拆除建筑分布 Reusable Spaces

■ 道路系统 Road System

■ 垃圾堆放位置 Garbage

■ 植被分布 Plants

■ 场地高程 Elevation　■ 汇水分析 Stream Analysis　■ 动物分布 Animals

教师评语：
村为原著，后有校来；村文化封闭没落但厚重绵长，校文化开放多元瞬息而变；村似敦厚长者，校如风华少年。校园建筑与古村建筑，风格、色彩、尺度、功能差异明显。什么样的空间能够促进两个群体的交流？选择什么样的物质载体实现文化认同？设计师选择中村民的生活质量修建共识的载体，组织修建、景观改造和水体净化工程为一体，一气呵成，系统集成本身具有一定的创新性，并实现了校园文化与村落文化共享、共建、共融的目标。

227

在研究范围内，尚存有兑山村的五个自然村。其中大部分仍然保留着祠堂和庙宇。围绕着这些堂庙，发展出了以水塘、广场、戏台、村公所、篮球场等要素组成的村中心空间。

There're five village communities remaining in the site, which are a part of Duishan administrative village. Most of them keeps an ancestral temple. The temples became the central element of the village center, with water ponds, squares, stages, administrative buildings and basketball grounds built around it gradually.

村落边界及村中心 Borders and Centers

		西珩	南尾井	宅内	下蔡		潘涂	
要素	祠堂	√	√	√	显明宫		常山宫	
	广场公共空间	√	√	√	√		√	
	戏台	√	√	√	√		√	
	村公社	√	×	×	×		×	
	篮球场	√	√	√	√		×	
	其他重点建筑	关帝庙	保获庙、安祠	土地庙	菜市场、洋楼		新厝祖厅、田仔乾祖厅	
	池塘	×	√	√	√		√	
	古树	古榕公园	龙眼林	古榕公园	√		古榕公园	
空间结构	通达性	良好	良好	较差	一般		良好	
	轴线关系是否明确	√	√	一般	√		有三条分散的轴线	
文化习俗	时间	正月初八	正月初八	正月初八	十月二十日	正月初八 七月初七	正月初八	三月初二 八月十六
	地点	兑山	兑山	兑山	兑山	显明宫	兑山	常山宫 常山宫
	节日及活动	游神	游神	游神	游神	供奉阎岁	游神	祭玄天上帝 祭赵子龙

村中心特性表 Characteristics of the Village Centers

西珩村中心 Xiheng Center

南尾井村中心 Nanweijing Center

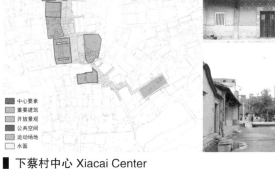

宅内村中心 Zhainei Center

下蔡村中心 Xiacai Center

下篇：介入与整合 —— 基于校村共生的华侨文化保护与侨村环境整治方案设计

调研分析
学生问卷调查 Student Questionnaire

村民访谈调查
Interviews with Villagers

策略与方案生成

策略生成
Strategy Formation

1
校村空间割裂
校园绿色廊道与村缺少联系

2
校园绿色廊道延伸
南北两村作为组团

3
校园绿色廊道向村内延伸
联系村内公共空间

实施过程设想
Implement Process

现状问题总结 Interviews with Villagers

图为兑山村内某连接建筑，村民私自搭建，结构不完整，无法使用，并且存在安全 隐患，需要拆除

图为兑山村内某处绿植，根据绿化现状分析可知村内有多处分散绿地，并且无人管理，杂草丛生，需要对其进行系统设计

图为村边污水涂鸦围墙，根据校村边界分析可知校村之间在空间割裂，需增加校村之间的联系

图为兑山村内一处池塘现状，可以看出池塘自身水质及与周边环境恶化状态恶劣，需要对其进行系统的净化处理

校村需求总结 Needs of Both Sides

池塘净化　　　　　　　　室内外活动空间
垃圾处理　　　　　　　　儿童活动空间
禽畜圈养
树木修整　　　　　　　　运动空间
　　　　　　　　　　　　阅览空间
废墟整理

项目清单
Project List

融合·共生与介入·整合 —— 华侨大学厦门校区与村落的城市设计研究

■ 整体平面图

针对校村空间割裂的问题，设计通过延伸校园现有绿色廊道网络进入现状村落区域的方式，整合村校空间，并以绿色廊道为切入点，改善村落环境。

■ Site Plan

In response to the spacial split between the campus and the villages, we designed an extension of the existing green corridor network of the campus, which will integrate the spaces, and will be a chance to improve the environment of the villages.

单体项目：华侨文化博物馆设计
Overseas Chinese culture Museum Design

本项目为华侨文化博物馆。出于博物馆本身的需要，展示空间应有很大的面积；然而巨大体量建筑的建设不符合介入的策略。因此从策展层面，我将华侨文化博物馆分为两部分：一是海外华侨文化展示，二是侨乡文化展示。海外华侨文化展示使用集中的空间，而侨乡文化展示可以依托村落，在村落内部展示。在这一策展方案中，博物馆变为一个主展馆和若干村内展厅，主展馆起到序厅和海外华侨文化展示的作用。本次设计着重进行了主展馆的设计。

The project is a museum of the culture of overseas Chinese. The project should be considered as a step of intervention, so the project can not be a big volume. So the exhibition is designed into two parts, one is about Chinese living overseas, while the other is about the homeland of overseas Chinese. Since the villages in the site are homeland of overseas Chinese, the latter can be distributed into the villages, and in this way the main hall can reduce its volume, as it only needs to contain the introduction lobby and half of the exhibitions. This project focuses on the main hall.

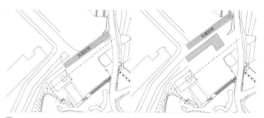

■ 边界处理 Boundary Treatment

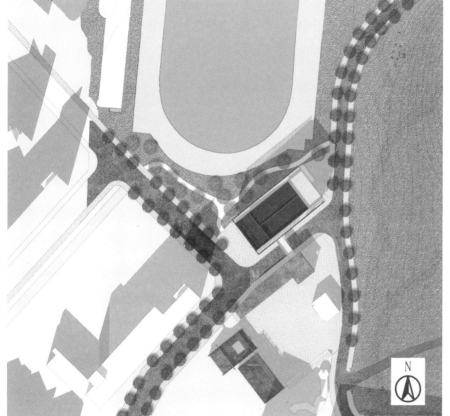

■ 总平面图 Site Plan

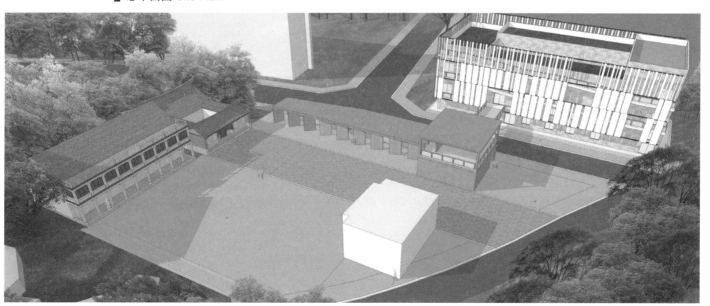

融合·共生与介入·整合 —— 华侨大学厦门校区与村落的城市设计研究

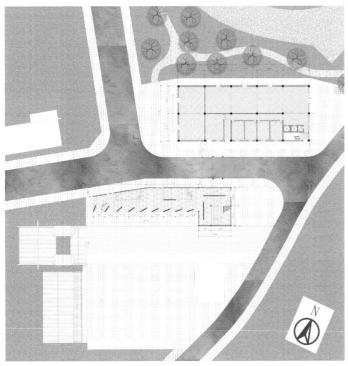

一层平面图 1st Floor Plan

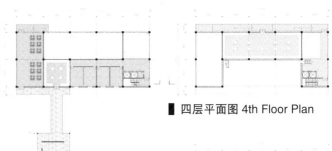

二层平面图 2nd Floor Plan

四层平面图 4th Floor Plan

三层平面图 3rd Floor Plan

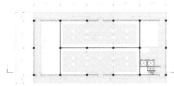

南立面图 North Elevation

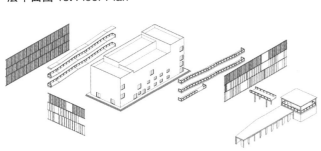

建筑要素构成分解 Building Composition

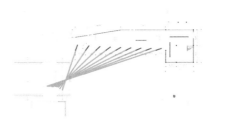

走廊视线分析 Sight Focus

参观流线 Visitor Route

单体项目：绿色空间 & 都市农业
Green Space & Urban Agriculture

本设计旨在呼应校园现有的行政轴线，建立一条文化轴线引入村内，并且作为村落绿色廊道的主要部分，将起到联系校园内绿色空间与村落绿色空间的作用。在绿色廊道上利用空置及废弃的古厝、宅基地、农田等，引入都市农业的功能，为侨村建立新的产业链，达到侨校与侨村的互惠共荣。

This design is intended to echo the existing campus administrative axis, to establish a cultural village as the main axis is introduced, and some villages of green corridors, green space and the village will play a main role in linking campus green space. In the green corridor, the use of the vacant and abandoned ancient houses, homestead, farmland and so on, the introduction of urban agriculture function, for overseas Chinese village to establish a new industry chain, to achieve the overseas Chinese School and overseas Chinese village mutual benefit and prosperity.

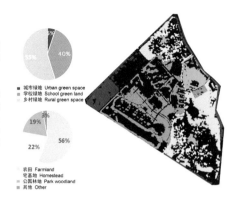

绿地分布 Distribution of Green Space

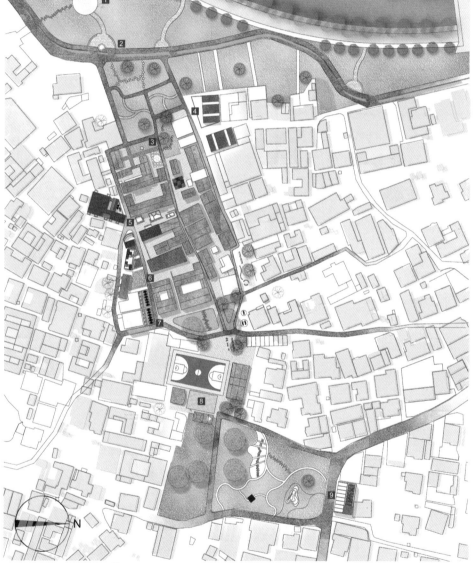

总平面图 General Layout

节点效果图 Node Renderings

场地现状 Site Situation

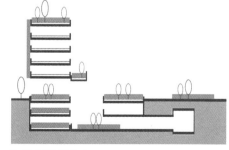

建筑种植分布 Planting in Architecture

融合·共生与介入·整合 —— 华侨大学厦门校区与村落的城市设计研究

空置古厝改造 Vacant Relic Renovation

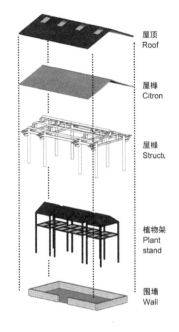

屋顶 Roof
屋椽 Citron
屋樑 Structu
植物架 Plant stand
围墙 Wall

古厝改造一层平面图 1st Floor Plan　　**古厝改造二层平面图 2nd Floor Plan**

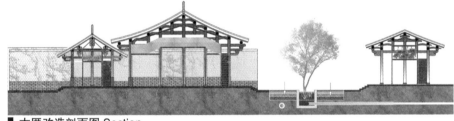

古厝改造剖面图 Section

农舍 Farmer Houses

空置宅基地改造 Vacant Homestead Renovation

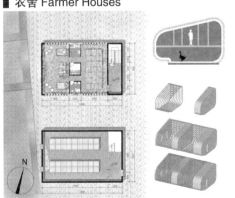

农舍平面图 Plan

空置宅基地一层平面图 1st Floor Plan

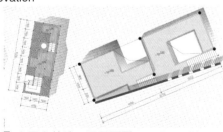

空置宅基地二层平面图 2nd Floor Plan

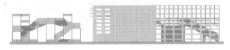

空置宅基地改造立面图 Elevation

单体项目：雨水花园活动中心设计

根据调研，拟为兑山村村民、华侨大学学生及儿童提供活动场所，并借此机会促进兑山村整体环境提升以及华侨大学与兑山村的交流融合。

利用池塘净化阶梯台地拓展为生态雨水花园，并将活动中心建筑置入其中，将室外活动场地与生态雨水花园结合设置，提高整体环境质量，使之成为绿色廊道上的重要景观生态节点。项目选址分别位于西珩、宅内、下蔡池塘处。

According to the research, I prepare to provide some activity places for villagers, HQU students and kids as well as to improve the environment of the village and the communication between village and school.

I transform the pond purification platform into a rain garden and add the activity center into it to combine the inside and the outside. The architectural design will not only improve the whole environment but also become one of the most important point on our green net.

The sites of the projects are in XiHeng, ZhaiNei and XiaCai.

选址分析

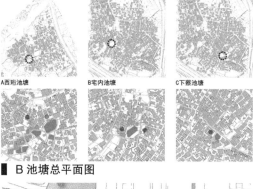

A 西珩池塘　　B 宅内池塘　　C 下蔡池塘

A 池塘总平面图 ／ B 池塘总平面图

A 池塘一层平面图 ／ A 池塘南立面图 ／ A 池塘剖面图

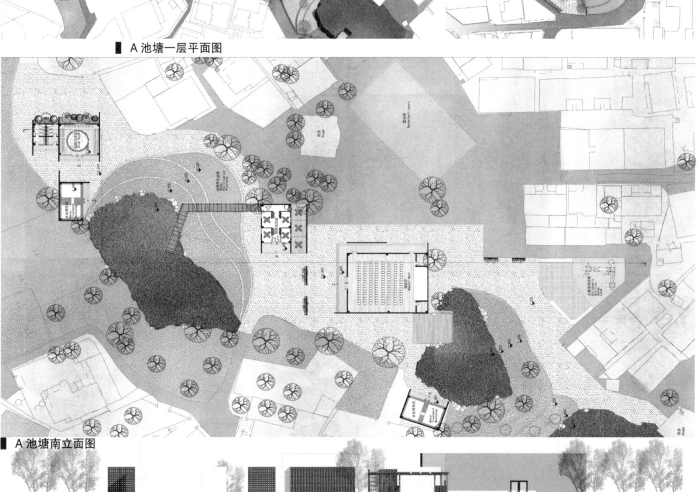

▎C 池塘一层平面图

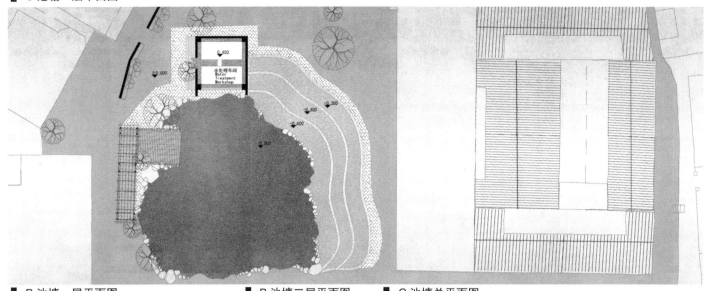

▎B 池塘一层平面图　　▎B 池塘二层平面图　　▎C 池塘总平面图

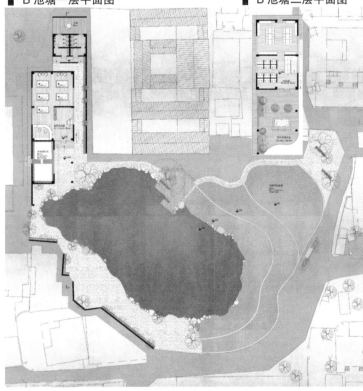

▎B 池塘南立面图

▎C 池塘南立面图

▎B 池塘剖面图

下篇：介入与整合——基于校村共生的华侨文化保护与侨村环境整治方案设计

华侨大学
HUAQIAO UNIVERSITY

目录 CATALOG

榕汇贯通 魏志明 吴佳贝 李炯 王迪
The Memory of Ficus Microcarpa

骑乐融融 任杰 徐嘉筠 孟娟 蒿倩
Riding to Promote the Harmony of School and Village

侨源共襄 张晗 吴平凡 杨琦辉 欧泊志
The Same Origin Share the Space

学生团队 Student Team

 张晗
 吴平凡
 杨琦辉
 欧泊志
 任杰
 李炯

 孟娟
 魏志明
 高倩
 徐嘉筠
 王迪
 吴佳贝

下篇：介入与整合 —— 基于校村共生的华侨文化保护与侨村环境整治方案设计

侨源 共聚

介入与整合：基于校村共生的华侨文化保护与侨村环境整治方案设计
Intervention and Integration: Preservation and Revitalization Design for the Traditional Village-within-Campus in HQU

侨源共聚
THE SAME ORIGIN SHARE THE SPACE

指导老师：刘塨 郑志 课程助教：姚敏峰
设计者：张晗 吴平凡 杨琦辉 欧泊志

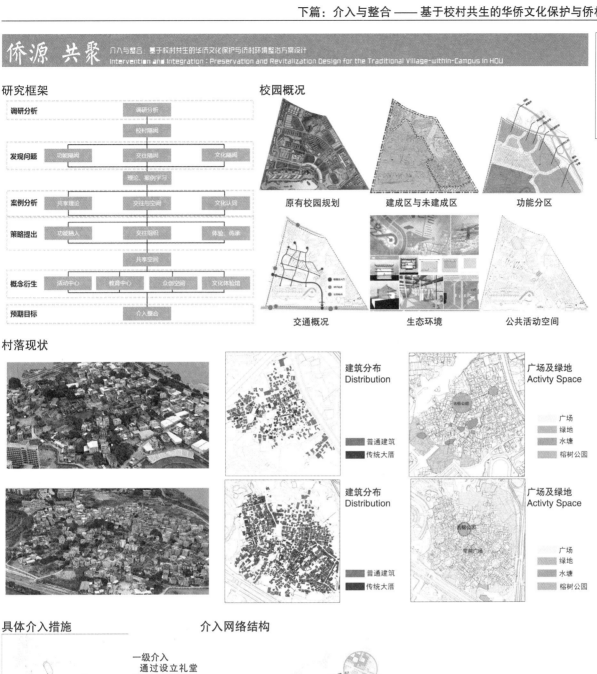

研究框架
调研分析 / 校村隔阂
发现问题：功能隔阂 / 交往隔阂 / 文化隔阂
理论、案例学习
案例分析：共享理论 / 交往与空间 / 文化认同
策略提出：功能植入 / 交往组织 / 体验、传承
共享空间
概念衍生：活动中心 / 教育中心 / 众创空间 / 文化体验馆
预期目标：介入整合

校园概况
原有校园规划 / 建成区与未建成区 / 功能分区
交通概况 / 生态环境 / 公共活动空间

村落现状
建筑分布 Distribution（普通建筑 / 传统大厝）
广场及绿地 Activity Space（广场 / 绿地 / 水塘 / 榕树公园）

具体介入措施

一级介入
通过设立礼堂和运动场地，在实现校村共享的同时，满足学校发展的刚需。

二级介入
对原本衰败的村落中心进行活化改造，改善场地环境，同时植入校园功能，实现共生。

三级介入
对村落中的建筑以及荒废场地进行改造，转换为校村共享活动场地，同时三级介入点也是作为串联一二级介入点的存在。

介入网络结构

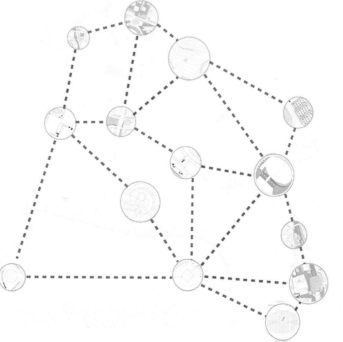

教师评语：

通过校园与村落在三种层次级别选点进行刚需设施建设、村落中心激活、废弃建筑整治，形成功能的共享与介入机制的支点。使校园与村落的日常活动各自延伸到"校-村"全域。通过社会生活的交融，激发带动空间秩序的交融。使融合不是一个结果，而是一个不断运动的过程。体现人居环境发展的特色。

融合·共生与介入·整合 —— 华侨大学厦门校区与村落的城市设计研究

侨源 共聚

介入与整合：基于校村共生的华侨文化保护与侨村环境整治方案设计
Intervention and Integration: Preservation and Revitalization Design for the Traditional Village-within-Campus in HQU

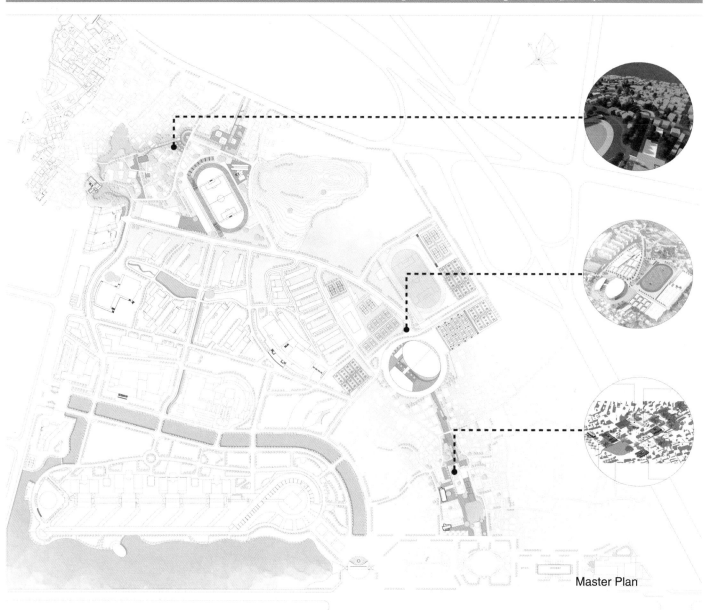

Master Plan

功能分区图　　道路系统图　　规划结构图

通过游神路线将校村紧密地联系在一起，通过文化达到整合介入点的效果

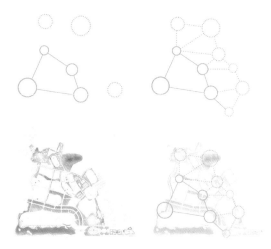

景观的延续

在改善村落原本衰败的环境以及对介入片区场所的营造后，实现了对校园原本景观的一个延续，使原本孤立的村落生态纳入到校园生态系统中来。实现了生态上的一种串联。

下篇：介入与整合 —— 基于校村共生的华侨文化保护与侨村环境整治方案设计

侨源 共聚

介入与整合：基于校村共生的华侨文化保护与侨村环境整治方案设计
Intervention and Integration: Preservation and Revitalization Design for the Traditional Village-within-Campus in HQU

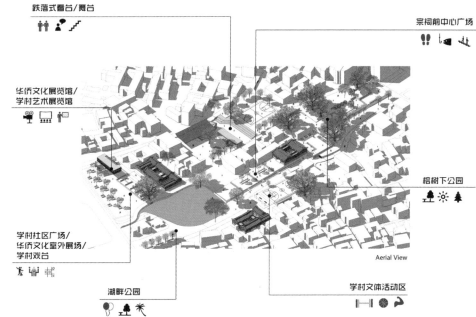

- 跌落式看台/舞台
- 宗祠前中心广场
- 华侨文化展览馆/学村艺术展览馆
- 榕树下公园
- 学村社区广场/华侨文化室外展场/学村戏台
- 湖畔公园
- 学村文体活动区

Aerial View

场地位于宅内古榕公园、宅内村中心祠堂附近，周边有一些极具价值的建筑以及生态景观，由于缺乏合理的引导，这些景观资源被逐渐埋没在杂乱无章的方块楼里。而改造策略则是通过对场地内的没有价值的建筑进行筛选拆除，让历史建筑充分呈现在人们面前，同时通过对景观的改造和开敞空间的营造，以达到活化村中心、促进村校交流、实现共享的目的。

First Floor Plan

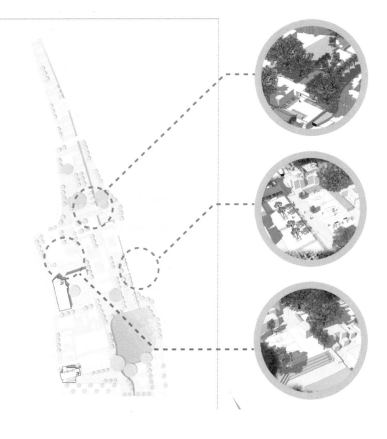

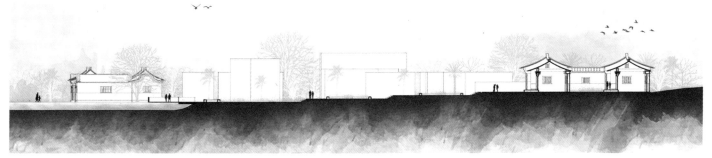

241

融合·共生与介入·整合——华侨大学厦门校区与村落的城市设计研究

侨源 共聚

介入与整合：基于校村共生的华侨文化保护与侨村环境整治方案设计
Intervention and Integration: Preservation and Revitalization Design for the Traditional Village-within-Campus in HQU

Base is located in the south of the village village village near the ancestral hall, the site is in front of the ancestral hall there is a building ruins. So the transformation strategy is to remove the ruins here, but the removal of the same time to retain a certain height of the wall as a future outdoor exhibition. For the treatment of the building, it is as much as possible to a layer of overhead, to ensure that students and villagers through the convenience of the same time to increase the more abundant space for people to stay and use.

Site Analysis

Block Analysis

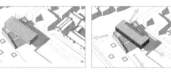

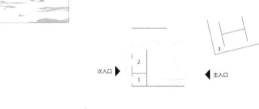

1 Canteen
2 Logistics Department
3 Outdoor Exhibition

Master Plan | Second Floor Plan | Underground Floor Plan

1-1 Section | South Elevation | East Elevation

侨源 共聚

介入与整合：基于校村共生的华侨文化保护与侨村环境整治方案设计
Intervention and Integration: Preservation and Revitalization Design for the Traditional Village-within-Campus in HQU

场地现状 Site Status

村中心祠堂 Ancestral Hall

废弃池塘 Untreated Pond

景观破败 Dilapidated Landscape

建筑废墟 Building Ruins

片区改造策略 Partition Design

校村现状

打通水系，置入共享空间

植入配套功能

片区改造分区 Partition Design

众创空间
滨水休闲广场
村落中心公园
运动驿站
村落文化艺术中心

在研究了该片区基地现状之后，总结出该基地原有村中心活力降低、景观破败、池塘无人治理、废墟未得到及时清理等问题。而我们需要做的是活化片区，使得村校使用者得到一个更好的共享空间，既有益于学校对村子的介入，又可以很好地整合村子现有的环境与空间资源。

在本方案中，我们从村子原有的池塘入手，将村落中几个池塘打通的同时，置入湖畔公园。然后结合村落中心这一较为虚化的意向，打造一个从校园视角可以直通村落中心的向心性广场空间。最后植入配套的功能和建筑。改善了村落环境的同时，为学校进一步发展提供了相应的物质空间的基础。

总平面图 Master Plan

改造前后肌理对比

改造前后平面对比

游神路线的影响

片区鸟瞰 Area Aerial View

村落文化艺术中心
村落中心广场
景观步道
休闲广场
运动驿站

局部表现 Local Perspective

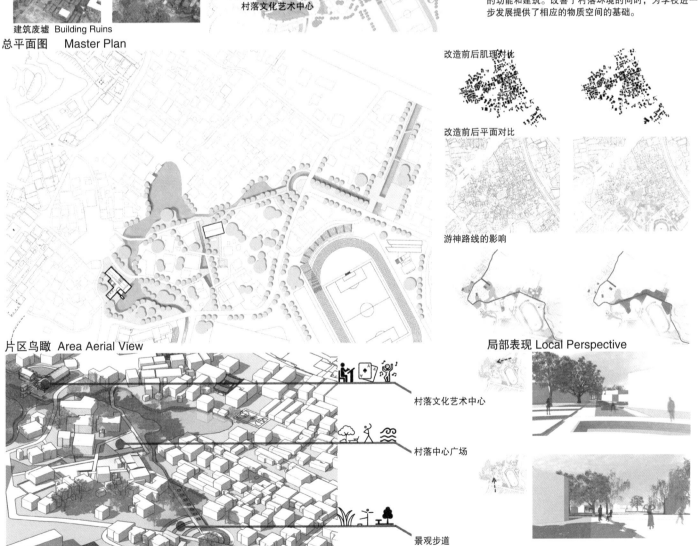

融合·共生与介入·整合 —— 华侨大学厦门校区与村落的城市设计研究

侨源 共聚
介入与整合：基于校村共生的华侨文化保护与侨村环境整治方案设计
Intervention and Integration: Preservation and Revitalization Design for the Traditional Village-within-Campus in HQU

考虑到该片区改造之后的各功能配比，发现该片区仍然缺少一个与学校功能相呼应的节点，需要寻找这样一个能起到兼顾校园生活与改造后片区功能需求的过渡。

In considering the functional proportions of the area after the transformation, we found that the area still lacks a node that echoes the school function, so we need to find a transition that can take on both the campus life and the functional requirements of the post.

主入口透视 Entrance Perspective

选址分析
Site Analysis

在我们重新规划设计的片区中，我们提出了将村中的古厝集中区改造为众创空间的想法。在空间联系上，校区内的大学生活动中心与众创空间只能隔着改造的村中心广场遥相呼应，这将带来些许的不便。所以我们在考虑基地选址的时候，主要考虑给来自校村的各股人流以更高的可达性，以提供更好、更高效地服务。

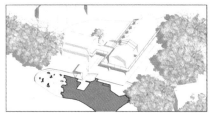

轴测图 Axonometry

功能分析
Functional Analysis

在这里我们需要什么？这个地块紧邻重新规划设计的湖畔公园，同时又地处西珩村与学校西街的连接点。这么好的区位优势，我们可以在这里尝试各种活动的发生。

设计说明
Design Description

本方案位于西珩片区，处于规划后湖畔公园旁，基地附近有保存完好的祠堂与广场。方案充分考虑周边的功能与建筑形式，定位为村落文化艺术中心。

在提取村落坡屋顶形式的同时，使用现代开窗与屋顶平台，并且局部一层架空，增加灰空间活动区域。建筑功能适应周边多样化人群的使用需求，形成多人群互动融合的多元化集散广场与空间。

体块生成
Block Generation

总平面图 Master Plan

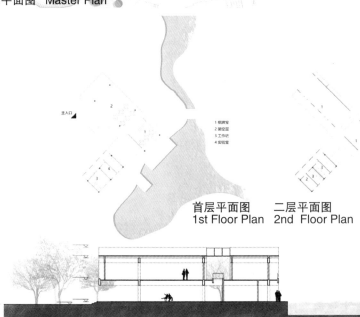

首层平面图　二层平面图
1st Floor Plan　2nd Floor Plan

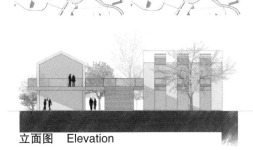

立面图　Elevation　　　剖面图　Section

下篇：介入与整合——基于校村共生的华侨文化保护与侨村环境整治方案设计

侨源 共聚
介入与整合：基于校村共生的华侨文化保护与侨村环境整治方案设计
Intervention and Integration: Preservation and Revitalization Design for the Traditional Village-within-Campus in HQU

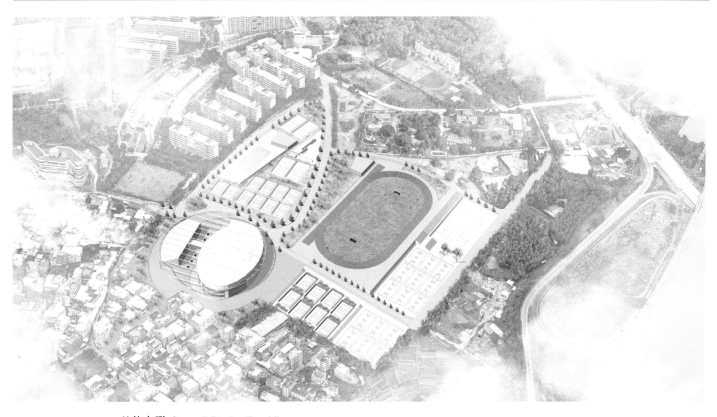

总体鸟瞰 Overall Bird's-Eye View

设置共享礼堂和共享会议中心
Set up shared auditorium and shared Conference Center

置入顶界面，形成灰空间
Insert the top interface and form the gray space

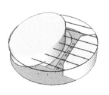

置入景观，活化空间
Implant landscape, activate space

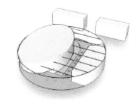

疏通景观轴，激活灰空间
Dredge the landscape axis and activate the gray space

现状"礼堂"-羽毛球馆
Status auditorium

原规划礼堂
Original planning auditorium

场地选址
Site selection

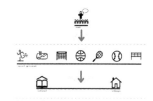

激活策略：在场地置入运动场地之后，通过礼堂的设立，汇聚引导人流，激活片区
Activation strategy: after the site is put into the sports ground, through the establishment of the auditorium, it will lead the crowd and activate the area

现状运动场地
Status sports field

原规划运动场地
Original planning sports field

礼堂选址
Auditorium site selection

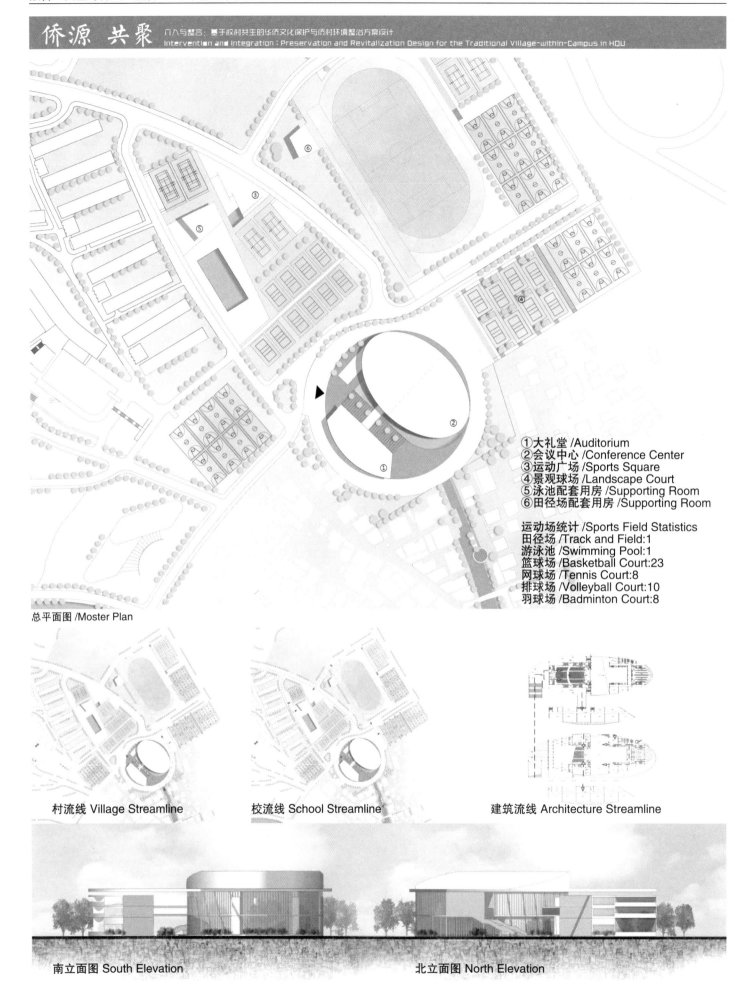

下篇：介入与整合 —— 基于校村共生的华侨文化保护与侨村环境整治方案设计

侨源 共聚

介入与整合：基于校村共生的华侨文化保护与侨村环境整治方案设计
Intervention and Integration: Preservation and Revitalization Design for the Traditional Village-within-Campus in HQU

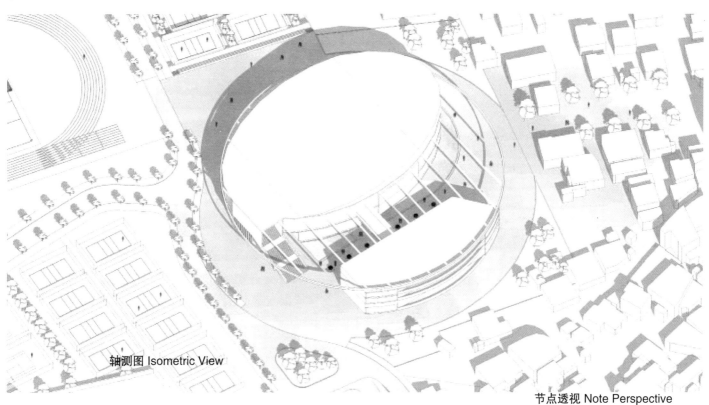

轴测图 Isometric View

节点透视 Note Perspective

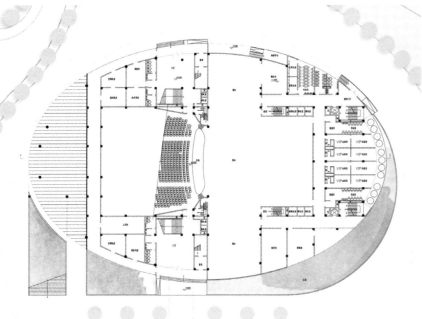

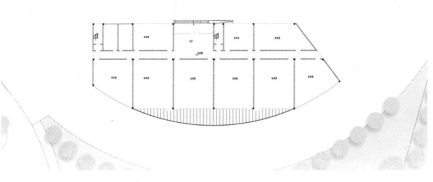

首层平面图 1st Floor Plan

247

侨源 共聚
介入与整合：基于校村共生的华侨文化保护与侨村环境整治方案设计
Intervention and Integration: Preservation and Revitalization Design for the Traditional Village-within-Campus in HQU

入口透视 Entrance Perspective

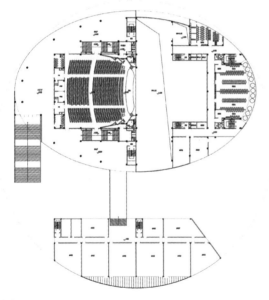

二层平面 2nd Floor Plan

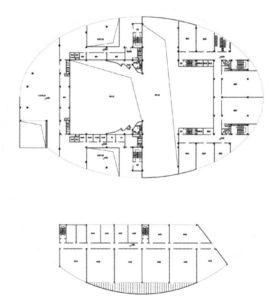

三层平面 3rd Floor Plan

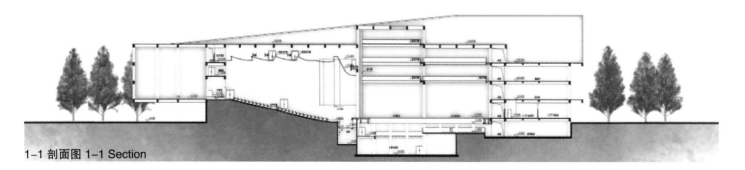

1-1 剖面图 1-1 Section

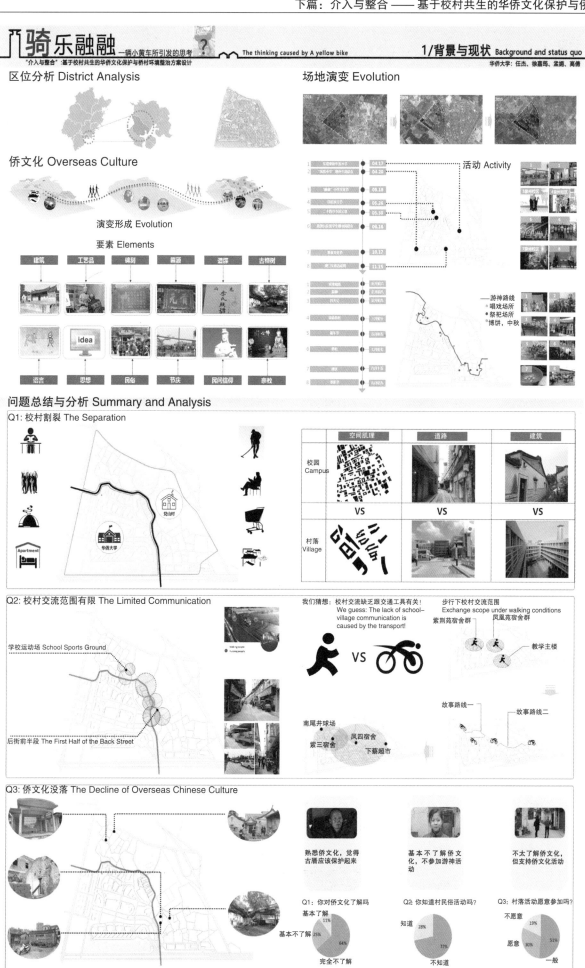

融合·共生与介入·整合 —— 华侨大学厦门校区与村落的城市设计研究

骑乐融融 —一辆小黄车所引发的思考
Thinking caused by A little yellow car

"介入与整合":基于校村共生的华侨文化保护与桥村环境整治方案设计

2/总体设计 Overall design

华侨大学:任杰、徐嘉筠、孟娟、高倩

概念框架 Conceptual Framework

| 思考 Thinking | 诊断 Diagnosis | 刺激 Stimulate | 编织 Weave | 补充 Supplement | 联动成面 Surface |

过程与目标 Process and Goal

主题阐释 Explanations of the Subject

骑 村落设置与学校道路相互串联的慢行系统,改善道路,提倡以"共享单车"为主的出行方式,方便出行,扩大校村交流圈!
The slow system between school and village improve the road, promote the convenient, travel way of ofo convenient, expand the exchange circle!

乐 整治村落环境,激活失落的公共空间,村落引入侨校活动,校侨体验侨村文化,使得侨校、侨村生活多姿多彩,和谐共生!
Through stimulating the lost public space, the life including overseas Chinese and overseas Chinese village become colorful!

融 通过公共空间的激活,侨文化活动的开展,慢行系统的串联,小黄车"共享单车"的触媒,最终校村融合发展,其乐融融!
Through the activation of public space, the development of overseas cultural activities and the slow system, school and village cheerfully development!

骑行系统构建 the Construction Riding System

现在人流集散 Current People Flow Distribution

骑行路线构建 Ride Route Construction

停车点布设 Parking Point Layout

侨文化元素提取 The Elements Extraction of Overseas Cultural

材料肌理 Material Texture | 界面构成 Interface Composition | 民俗风情 Customs | 色彩 Color

传统形制转型 1 The Transformation Form One | 传统形制转型 2 The Transformation Form Two | 传统形制转型 3 The Transformation Form Three

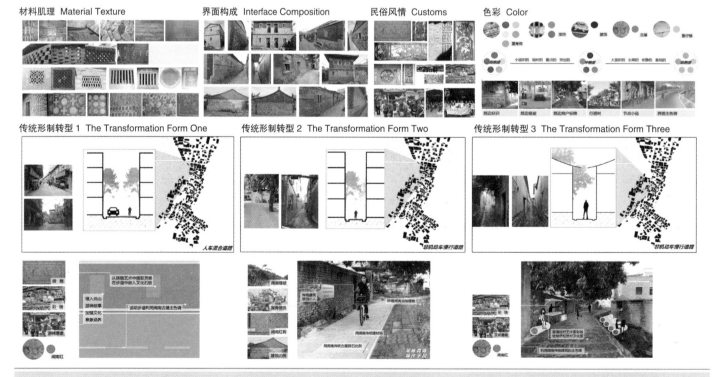

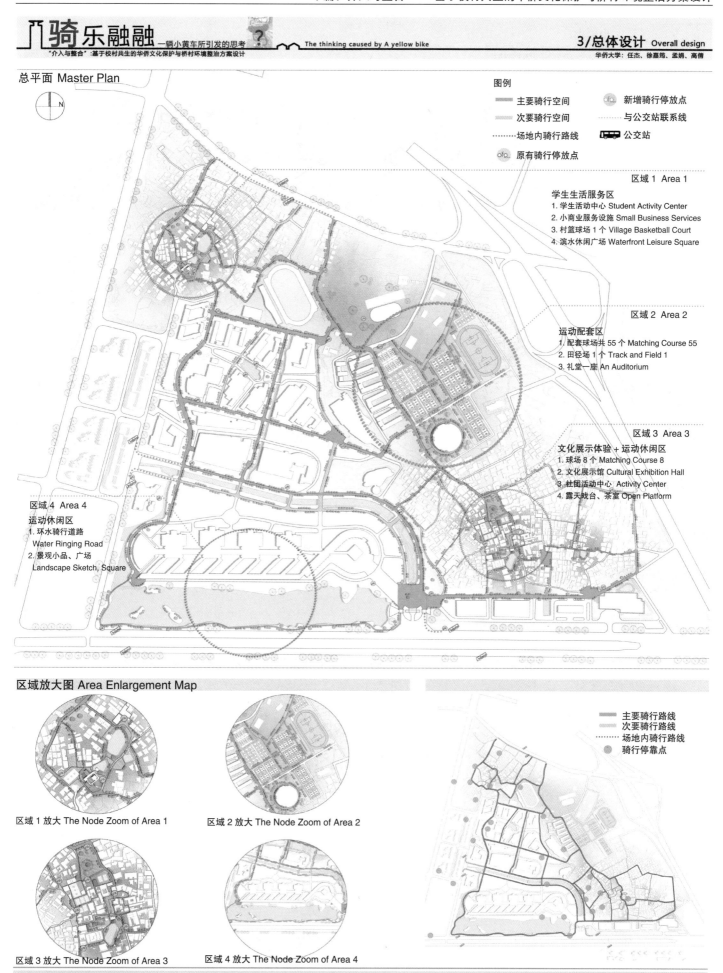

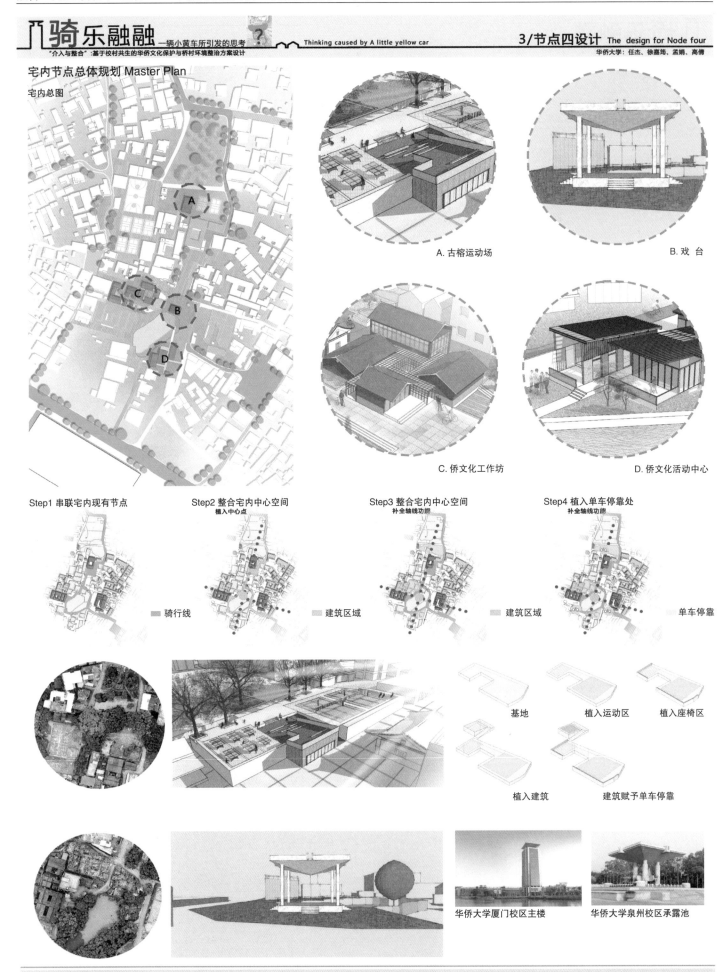

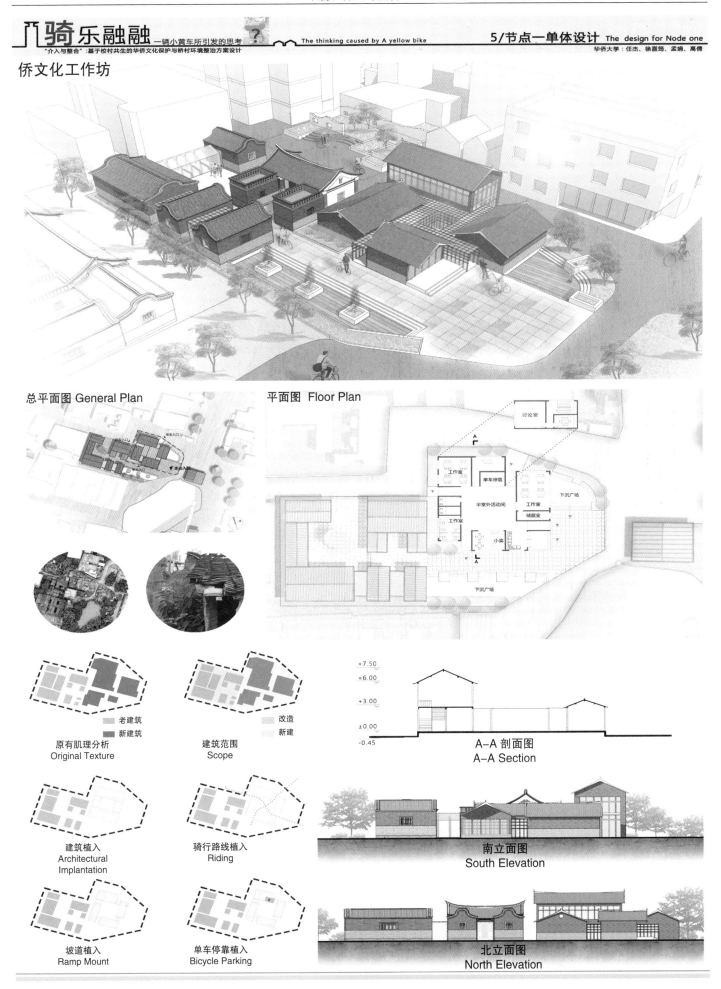

融合·共生与介入·整合 —— 华侨大学厦门校区与村落的城市设计研究

骑乐融融 —— 一辆小黄车所引发的思考
The thinking caused by A yellow bike

"介入与整合":基于校村共生的华侨文化保护与桥村环境整治方案设计

6/节点一单体设计 The design for Node one

华侨大学：任杰、徐嘉筠、孟娟、高倩

侨文化活动中心

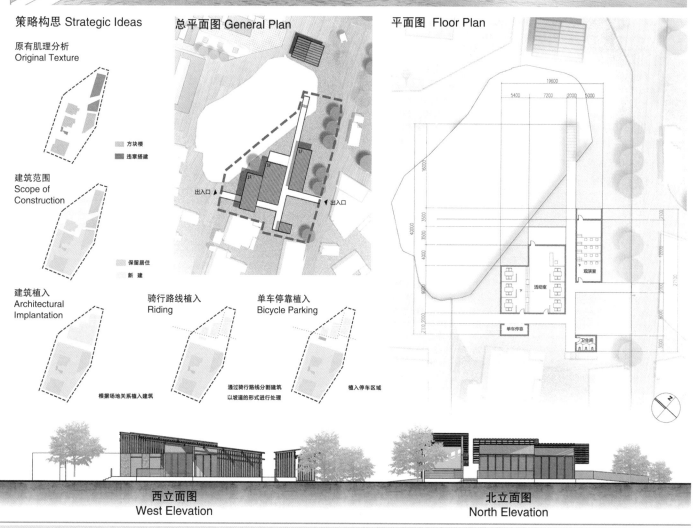

下篇：介入与整合 —— 基于校村共生的华侨文化保护与侨村环境整治方案设计

骑乐融融 —— 一辆小黄车所引发的思考

The thinking caused by A yellow bike

"介入与整合"：基于校村共生的华侨文化保护与桥村环境整治方案设计

7 / 节点二设计 The design for Node two

华侨大学：任杰、徐嘉筠、孟婧、高倩

节点效果图 Node Effect Graph

锻炼篇 (Exercise articles):
骑行道路完善之后，已经喜欢上了每天骑行半小时
After the ride is perfect, already Like riding a day for half an hour

休闲篇 (Leisure articles):
骑行于村，偶尔碰到一场激烈的篮球赛，很过瘾
Riding in the village, occasionally encounter a fieldIntense basketball game, very fun

生活篇 (Life articles):
骑累了，停下来，于榕树下纳凉，聊天，也十分惬意
Ride tired, stop, under the banyan tree Cool, chat, but also very comfortable

Elevation

Streamline
- Accessible routes
- ▲ Entrances and exits

Building Quality
- Good
- General
- Poor

Green Space
- Gurong Park
- Square
- Waste green land
- Old trees

平面图 Plan

- 视角2 骑行停靠点 Riding a stop
- 社区活动中心 Community Center
- 视角5 骑行停靠点 Riding a stop
- 图书室 bathroom
- 篮球场 Basketball court
- 视角4 榕树下空间 Under the banyan tree space
- 戏台 Stage
- 视角1 休憩石凳 Resting stone bench
- 视角3 羽毛球场 badminton court
- 骑行停靠点 Riding a stop
- 主要骑行路线 Main ride route
- 次要骑行路线 Minor ride route

策略构思 Strategic Ideas

1. 梳理流线
2. 新增骑行停靠点
3. 运动场地规划
4. 运动休闲配套
5. 榕树空间整治
6. 空间渗透与激活

节点透视 Node Perspective

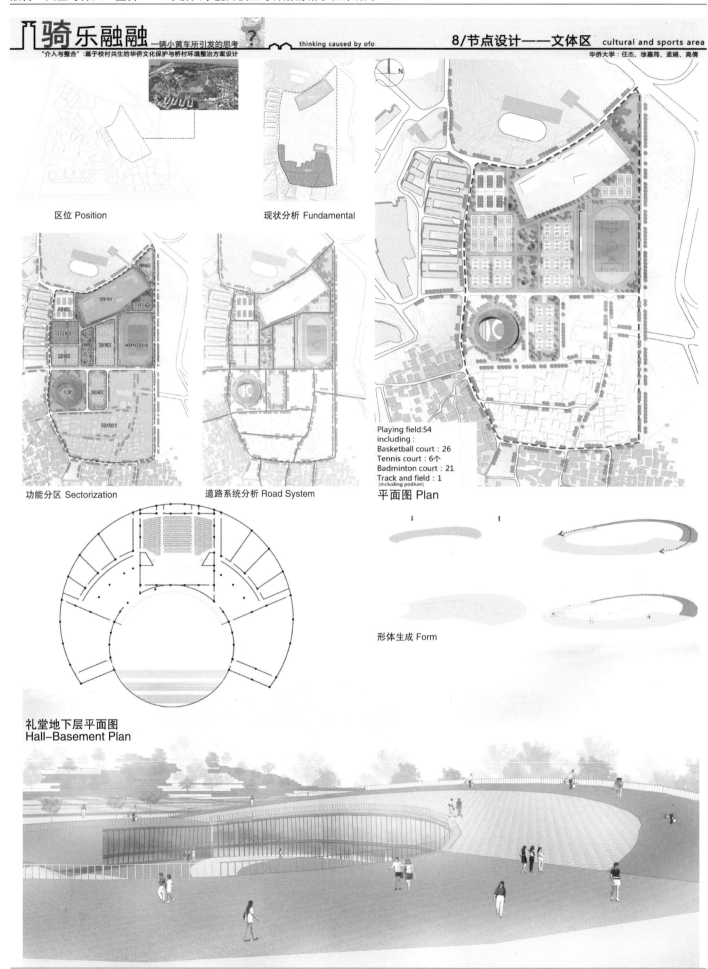

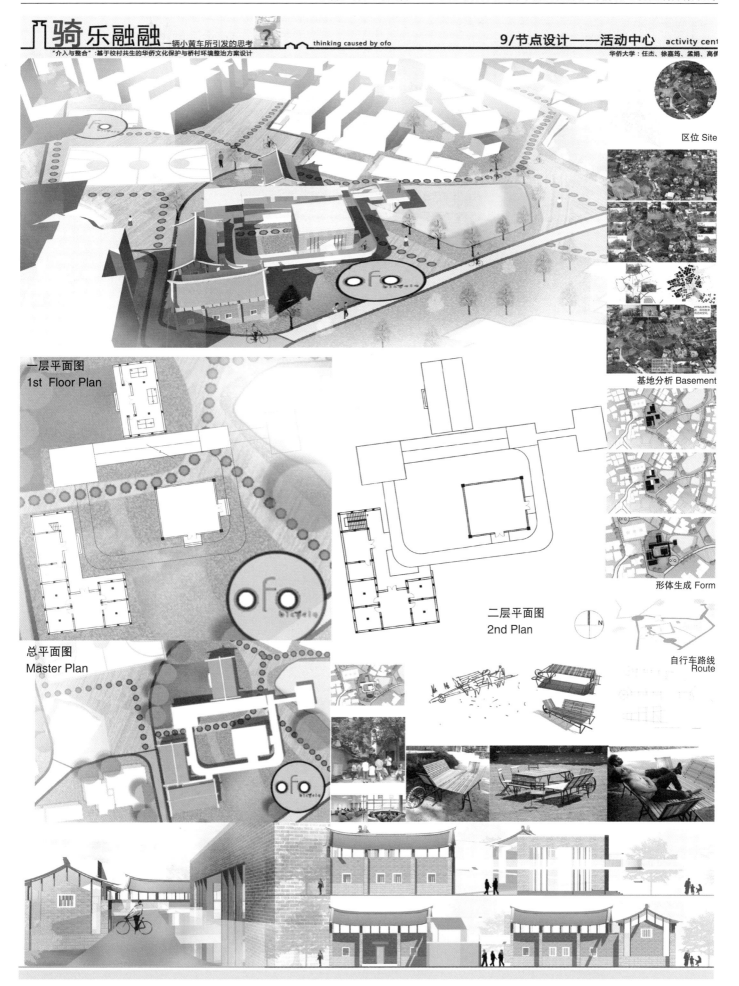

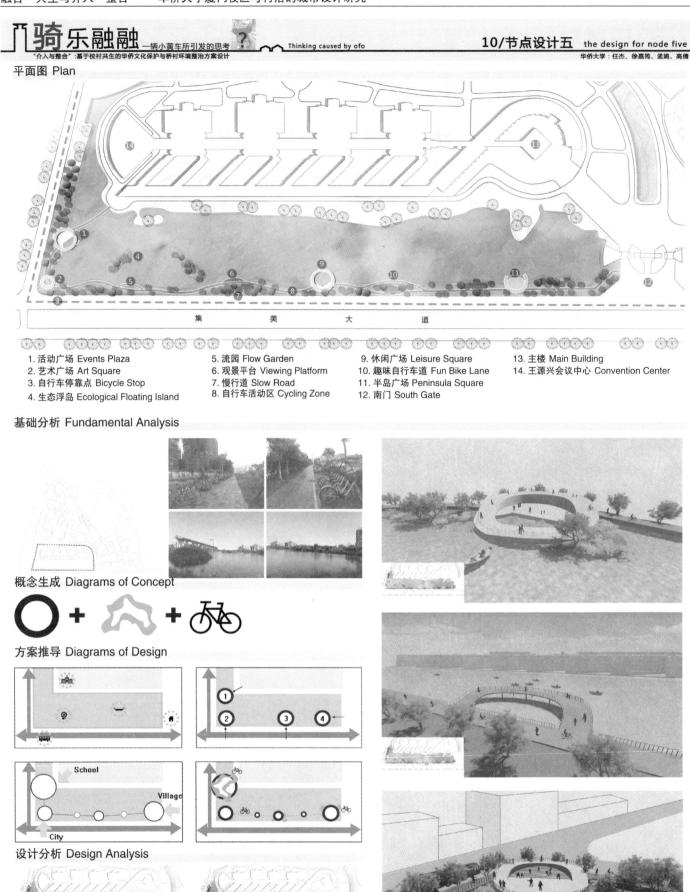

榕滙貫通
THE MEMORY OF FICUS MICROCARPA
介入与整合：基于校村共生的华侨文化保护与侨村环境整治方案设计

课题背景与现状分析 01
BACKGROUND AND STATUS ANALYSIS

校村背景 / BACKGROUND

Step.1 侨村背景

乡土意识 Local consciousness

家族观念 Concept of family

民间信仰 Folk religion

一方面，移民的方式以邻里亲友牵引出国为主，并形成了同乡邻里聚居于国外某一地区的普遍现象。另一方面，在中华文化重教化的思想影响下重视教育，怀着"教育救国"的思想纷纷回乡兴办教育事业。

华侨华人对家庭和集体主义的重视，源于中国传统文化中的亲缘关系，而侨村出现的侨汇和侨房可以看作是这些观念作用于侨乡社会和华侨华人社会的直接后果。

一般家庭大厅后壁正中均设祖先神位或遗像。逢年过节和每月初一、十五，均要祭拜祖先，其中以元宵、中元、冬至、除夕四节最为隆重。

Step.2 侨校背景

二十四节令鼓　赛龙舟　东南亚"泼水节"　中华文化大乐园　境外生"美食街"

Step.3 校村共融

华侨大学　经济　教育　服务　社区　交流　环境　活动

古榕公园-交流场所　原住民　关帝庙-文化记忆　村中心、宗祠、古厝、戏台-传承　榕树-村落象征

现状调研 / STATUS

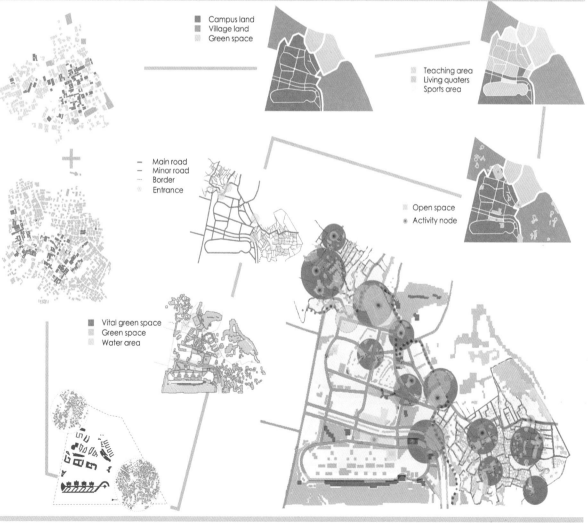

- Campus land
- Village land
- Green space

- Teaching area
- Living quaters
- Sports area

- Main road
- Minor road
- Border
- Entrance

- Open space
- Activity node

- Vital green space
- Green space
- Water area

设计者：魏志明　吴佳贝　李炯　王迪
指导老师：刘塨　郑志
课程助教：姚敏峰　赖世贤

榕滙貫通
THE MEMORY OF FICUS MICROCARPA

教师评语：

作品抓住榕树对于闽南地域文化，特别是华侨文化的独特含义，通过榕树空间的发掘、整治、拓展与激活，串联起整个场地的华侨文化元素，先把两个相互隔离的村落连在一起，然后把校园与村落连成一个整体。形成师生与村民共同生活的网络。这里校的华侨文化与侨村的华侨文化得到充分的共同体现。

下篇：介入与整合——基于校村共生的华侨文化保护与侨村环境整治方案设计

融合·共生与介入·整合——华侨大学厦门校区与村落的城市设计研究

榕汇贯通
THE MEMORY OF FICUS MICROCARPA
介入与整合：基于校村共生的华侨文化保护与侨村环境整治方案设计

改造策略与具体方案
STRATEGY AND PROPOSAL　02

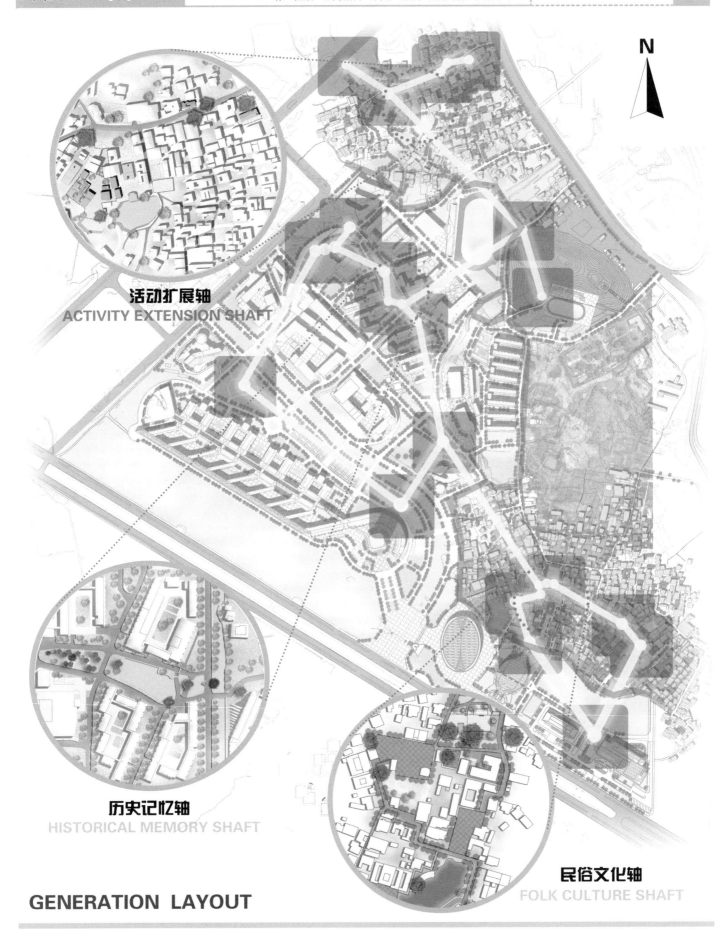

活动扩展轴　ACTIVITY EXTENSION SHAFT

历史记忆轴　HISTORICAL MEMORY SHAFT

民俗文化轴　FOLK CULTURE SHAFT

GENERATION LAYOUT

下篇：介入与整合——基于校村共生的华侨文化保护与侨村环境整治方案设计

榕滙貫通 THE MEMORY OF FICUS MICROCARPA
介入与整合：基于校村共生的华侨文化保护与侨村环境整治方案设计

改造策略与具体方案 / STRATEGY AND PROPOSAL 03

鸟瞰图 / Aerial View

总平面 / Master Plan

要素轴测 / Element Axonometry

植入建筑
Implanted in the building

榕树、植被
Banyan, vegetation

周边建筑、古厝
Surrounding buildings ancient houses

绿地、绿化
Green, green

地块、道路、水面
Plots, roads, water

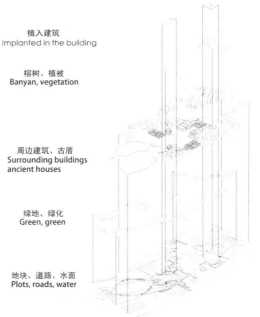

体量生成 / Volume Generation

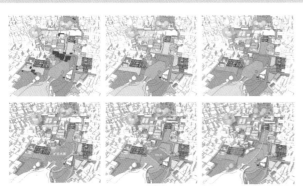

空间轮廓 / Spatial Contours

融合·共生与介入·整合 —— 华侨大学厦门校区与村落的城市设计研究

榕滙贯通 THE MEMORY OF FICUS MICROCARPA
介入与整合：基于校村共生的华侨文化保护与侨村环境整治方案设计

改造策略与具体方案 STRATEGY AND PROPOSAL　04

文化中心鸟瞰 /Cultural Center Aerial View

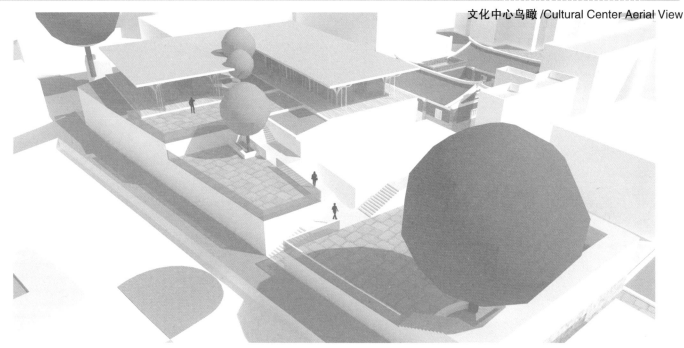

总平面 /Master Plan

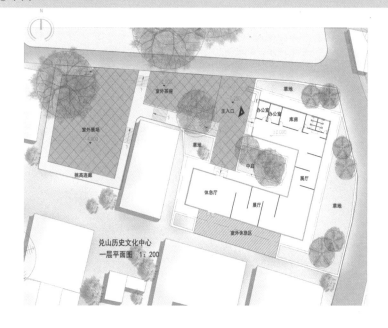

广场、绿地、人流 /Square, Green, Flow

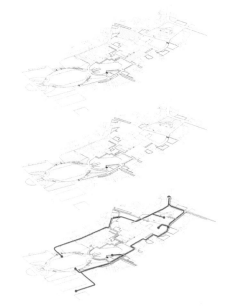

体量生成 /Volume Generation

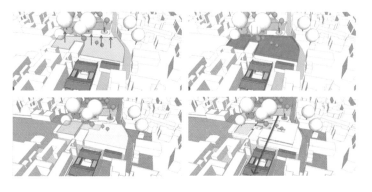

综合演出舞台 /Body Block Generation

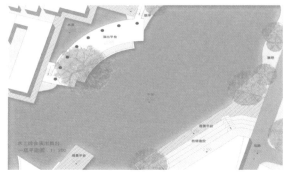

下篇：介入与整合——基于校村共生的华侨文化保护与侨村环境整治方案设计

榕滙贯通
THE MEMORY OF FICUS MICROCARPA
介入与整合：基于校村共生的华侨文化保护与侨村环境整治方案设计

改造策略与具体方案 05
STRATEGY AND PROPOSAL

鸟瞰图 /Aerial View

总平面 /Master Plan

概念演绎 /Inference

将榕树串联，强化榕树之间的联系，赋予树下空间职能，引导村民进校，与学生共同享受校园资源，为校村融合奠定基础。

Series of banyan tree, Strengthen the connection between trees.The function of space under the banyan tree.Guide the villagers into the school.

场地分析 /Location Analysis

体量生成 /Volume Generation

| 原始绿地 | 扩宽水面 | 榕汇贯通 |
| 水塘封闭 | 挖地造湖 | 规划道路 |

小桥　凌波微起　　座椅　环水而生　　场所　聚集人流

融合・共生与介入・整合 —— 华侨大学厦门校区与村落的城市设计研究

榕滙贯通 THE MEMORY OF FICUS MICROCARPA
介入与整合：基于校村共生的华侨文化保护与侨村环境整治方案设计

改造策略与具体方案 06
STRATEGY AND PROPOSAL

鸟瞰图 /Aerial View

总平面 /Master Plan

概念演绎 /Inference

由"绿树成荫"这个成语引发的灵感，通过新栽种榕树的方式来打造一个树阵广场，供村民与学生在广场内遮阳聊天，串联榕树，激活公共空间。

Triggered by this idiom tree-lined inspiration, through the banyan way to build a new plant trees, square, for the villagers to chat with students in the square sunshade, tandem banyan, activate the public space.

体量生成 /Volume Generation

场地分析 /Location Analysis

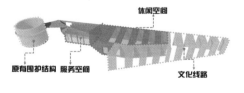

下篇：介入与整合——基于校村共生的华侨文化保护与侨村环境整治方案设计

榕汇贯通
THE MEMORY OF FICUS MICROCARPA
介入与整合：基于校村共生的华侨文化保护与侨村环境整治方案设计

改造策略与具体方案 | 07
STRATEGY AND PROPOSAL

概念演绎 /Inference

总平面 /Master Plan

Taking the existing pottery art manual creation workshop as the starting point, combined with the banyan space to design, the internal layout of a number of workshop units.

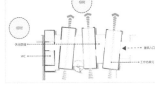 

场地分析 /Location Analysis

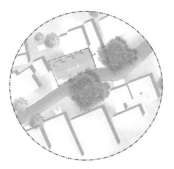 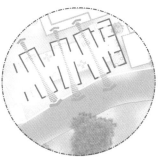  

融合·共生与介入·整合 —— 华侨大学厦门校区与村落的城市设计研究

榕滙貫通 THE MEMORY OF FICUS MICROCARPA
介入与整合：基于校村共生的华侨文化保护与侨村环境整治方案设计

改造策略与具体方案 08
STRATEGY AND PROPOSAL

总平面 /Master Plan

场地分析 /Location Analysis

新建筑在旧铁皮墙的基础上，将其面对榕树部分改造成大面积的玻璃幕墙，结合坡屋顶并保存原有的结构框架体系。建筑在入口处退出一小部分共享空间，在内部考虑榕树景观点，做成阶梯状看台，并在屋顶开设天窗，营造良好的阅读环境。

一层：儿童书院、休闲茶座
二层：侨文化展示

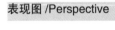

表现图 /Perspective

建筑立面 /Elevation

2017 国际五校研究生联合设计过程回顾
2017 International Joint Studio for Five Universities Review

开幕讲座
Opening Lectures

现场调研
Site Investigation

小组讨论
Group Discussion

调研汇报
Investigation Report

下篇：介入与整合 —— 基于校村共生的华侨文化保护与侨村环境整治方案设计

视频课程
Video Courses

答辩
Reply

相关学术讲座
The Academic Forum

刘塨 教授

讲座题目
《"介入与整合"——基于校村共生的华侨文化保护与校村环境整治》

许金顶 教授

讲座题目
《闽南华侨历史文化与兑山村变迁简史》

边经卫 教授

讲座题目
《基于"共同缔造"理念下的规划方法建构》

许懋彦 教授

讲座题目
《记忆的风景——欧洲城市公共空间奖及作品选讲》

陈志宏 教授

讲座题目
《结合民间史料的传统村落景观研究与保护》

赵燕菁 教授

讲座题目
《边缘地区城市化途径》

张路峰 教授

讲座题目
《建筑介入城市 设计改变生活》

2017 国际五校研究生联合设计教学总结与反思
2017 International Joint Studio for Five Universities Teaching

一、题目及背景

本次题目是探讨厦门校区空间（华侨大学）与周边社会空间的关联性的课题。

1. 大学校园作为教育空间，通常被认为是自我封闭、自我完结的空间，但是实际上学生的日常生活并没有在校园内部完结，反之会更多地与周边的居民、商店街等发生关联。比起大包大揽地整合校园与周边区域的关系（即二者择其一，有"一个吞噬另一个"的可能性），更重要的是如何让两者和谐共存，互利互惠。

2. 提起校园，一般都包含了大尺度的建筑（例如教学楼、体育馆），这与普通居民生活所在的人体尺度的住宅有很大差异。如何调和两者是亟须解决的问题。关于尺度，与整体规划相对应的是自然生成的居民生活空间（经过长期的点滴的积累而成，并将持续变化）。这一固定与流动的两者组合也是不得不考虑的问题。在这一层面，构成形态上将会很有趣。

3. 除了空间性之外，这一区域也具有历史性的特性。这里到处残留了或新或旧的历史纪念物；同时，这里也有或共有或私有的所属不同。如何把这些不同属性归类活化也是需要着手的问题。

二、学生作品

尽管调研周期短暂，但是同学们在现场用自己的脚步丈量、发现，通过这样细致缜密的调查一定会有很多感受和感想。

1. 同学们的提案中，有统筹校园和周边住区的整体规划方案（2016 清华大学方案）：通过圆环的形态连接校园与周边街区，并不单纯针对其中一者而是兼备考虑两者如何结合。比起桥状连接的方法，这种环状连接的方法我认为更有效。

2. 相较上述的大尺度方案，也有通过点状方式连接各要素，更接近人体尺度，更贴合当地地形脉络的提案（2016 早大方案）。细想一下，各个要素点状地、缓慢地呈现出来，最后像项链一般串联成一个整体。与上一方案不同，尽管看上去是分散的，不太显眼甚至有些弱势，但这样的连接方法也是同样甚至更有效的。

3. 当然其中有各种各样的方案，有强势的有弱势的，有看上去一目了然的，也有忽隐忽现的。可以说这些提案都是非常有效且具有多样性的。

三、WORKSHOP 的目的

1. 说起 WORKSHOP 的目的，举个例子。比如平常看惯的事物风景，在不同地方到来的陌生人（Stranger）看来就会看到理所当然之外的层面。例如对于场地，华侨大学的同学们当然是司空见惯的，但在中国其他地区、日本东京，甚至其他国家到来的同学们看来，一定会有不同的见解。

2. 反而言之，作为陌生人如果只是不明所以地走马观花，就只能看到表面的现象，会有太过强调视觉形态的倾向。但如果能和对此区域熟悉的人一起边走边看，就能了解其背后深层的机制，眼中所关注的对象也会发生变化。

3. 而在我看来，WORKSHOP 本身的意义不就在于此，不就在于协同不同背景的人，一起调研，共同发现，协同设计，和谐共生吗？

早稻田大学　古谷诚章
2018 年 5 月

后记
Postscript

2016-2017年两届联合设计的成果终于出版了,导师组成员在此向大家致谢。

首先感谢参加本课程学习的学生们,正是你们在学习研究过程中的辛勤努力,让本书具备了一定的学术价值。同时,感谢拨冗为本课程开出讲座的教授们,让本课程的研究站在一个较高层次的学术平台。

此外,还要感谢华侨大学建筑学院党政领导、团委老师们在课题调研过程中对解决数十位外地学生住宿及调研时繁杂问题所付出的辛苦;感谢华侨大学研究生院、国际交流合作处、信息化建设与管理处的领导和老师们所给予的支持与帮助。

最后,感谢组织本书排版的赖世贤老师和参与排版的张晗等2017级研究生们。

2018年新课程移师特区澳门,新的努力已经开始,期待见到更好的成果。

五校联合设计教学团队

2018年5月